THE RIGHT TO THE CITY

The Right to the City

Social Justice and the Fight for Public Space

DON MITCHELL

THE GUILFORD PRESS
New York   London

© 2003 The Guilford Press
A Division of Guilford Publications, Inc.
72 Spring Street, New York, NY 10012
www.guilford.com

Postscript © 2014 The Guilford Press

Printed in the United States of America

This book is printed on acid-free paper.

Last digit is print number: 9   8   7

Library of Congress Cataloging-in-Publication Data

Mitchell, Don, 1961–
   The right to the city : social justice and the fight for public space
   / Don Mitchell.
         p. cm.
Includes bibliographical references and index.
   ISBN 978-1-57230-847-3 (pbk. : alk. paper)
   1. Public spaces—United States. 2. Social justice—United States.
3. Protest movements—United States. 4. Homeless persons—Civil
rights—United States. 5. Urban geography—United States. I. Title.
   HT123 .M5656 2003
   307.76'0973—dc21
                                                                   2002155032

Acknowledgments

Prompted by my own political involvements and a task set for me by Bob Lake in his seminar on locational conflict, I have been thinking and writing about public space for more than 10 years now. I've also been piling up debt upon intellectual debt as my ideas have developed and transformed. I can't begin to thank everyone who has had a hand (wittingly or not) in this book, but here is a start. Thanks to:

Bob Lake for helping spur this interest in the first place and for continuing to ask extremely productive and insightful questions; and Neil Smith for encouraging my work early on and shaping it in more ways than I probably know.

All the reviewers—anonymous and revealed—who suffered through early drafts of these chapters as they worked their way into print as journal articles, especially Sallie Marston, who was an early, demanding, and very supportive reviewer, and who remains so; and Nick Blomley, who has read and supportively critiqued almost everything I have written on public space and on law, and who in addition read through and manuscript.

Matt Hannah, who also read the whole manuscript and provided a deeply insightful commentary, some of which inspired changes in the volume, but much of which I hope will bear fruit later, not only in my own work but in Matt's too.

Rich Van Deusen and Clayton Rosati, who read through the whole manuscript in an effort—perhaps in vain—to improve my English and to keep me from making a fool of myself (Rich additionally tracked down illustrations); and Reecia Orzeck, who tracked down last-minute references. All have become comrades and have made the People's Palace a great place to be.

Lynn Staeheli, who as a collaborator and friend has pushed my ideas in new directions, and who has only occasionally complained about the research wine I spilled all over her in San Diego.

Peter Wissoker, who commissioned a book from me sometime last century and who left The Guilford Press before seeing the result—despite years of advice, gentle prodding, and coffee; Kristal Hawkins, who has taken over for Peter, and figured out how to turn up the heat enough so I actually did finish the book; K. K. Waering, Jr., who did an outstanding job of copyediting; and Anna Nelson, who saw the book through production.

My parents, Jim and Bunny, and my brother, David, and his wife, Nora, who all provided information and collected newspapers for me during the 1991 Berkeley riots and who then took the time to disagree with my interpretations.

Ann and Bob Millar, who have once more provided a genial and comfortable place in the west of Scotland to put the final touches on a book.

The geography departments at Glasgow University and the University of Oslo, which provided both resources and comradeship when I was in the last stages of writing this book.

And, especially, Susan Millar, who has questioned me, supported my work, and kept me sane in the midst of the madness that is our everyday lives.

Contents

The *right to the city* is like a cry and a demand.
—HENRI LEFEBVRE,
"The Right to the City" (1996 [1968])

The universal consequence of the crusade to secure
the city is the destruction of any truly democratic space.
—MIKE DAVIS, "Fortress Los Angeles" (1992)

The Fight for Public Space

What Has Changed?

On the Sunday following the horrific terrorist plane crash attacks of Tuesday, September 11, 2001, the *New York Times* ran a full-page feature story asking what it would take to make New York's "public spaces safe from attack" (Barstow 2001, 1:16). The *Times* interviewed "nine security and terrorism experts" to "envision a New York City of maximum security where money was no object in the pursuit of safety." The discussion was compelling. "Security options once dismissed as unpalatable, impractical or too expensive would be embraced," the *Times* wrote. "There would be long lines and intrusive and random searches, new identification systems and a strange new vocabulary of terms like biometrics, bollards, bomb mitigation containers and smart doors." One of the experts said, plainly, that "You would have to develop a fortress mind-set" (quoted in Barstow 2001, 1:16).

To some extent, New Yorkers have been preparing for that mind-set for quite some time. Well before September 11, public space had already been significantly fortified—or at least radically transformed—in the name of security over the past generation. Parks had been reconstructed and fenced, and special enclosed areas for children and their guardians had been established. The policing of public spaces ranging in size from small squares to fairly large urban parks and train stations had been turned over to private police forces paid for by, and under the direction of, Business Improvement Districts. New strictures on behavior had become not only commonplace but also expected (and always indicated by prominent signs) in the city's streets. Surveillance cameras had become an everyday part of the landscape. Whole public spaces had been closed off for much of the day, locked tight against unwanted users.

The context for these transformations in New York, as in most American cities, of course, was not the threat of terrorist attack but rather the fear of inappropriate users: the homeless, drug dealers, loitering youth—and, not inconsequentially, political activists protesting in front of city hall, marching in the streets, or rallying in parks and squares. The solution to the perceived ills of urban public spaces over the past generation has been a combination of environmental change, behavior modification, and stringent policing. The putative reason is to assure that public spaces remain "public" rather than hijacked by undesirable users.

The experts the *Times* interviewed in the wake of the terrorist attacks all agreed that further environmental modification was now necessary, including the closing off of the steps to churches, cathedrals, and synagogues, the installation of hundreds of surveillance cameras around important public spaces and along "vulnerable" streets, the installation of more "bomb-proof" windows, trash cans, and so forth. Policing too, they agreed, should be stepped up. Some argued for the deployment of armies of bomb-sniffing dogs and their handlers, even the authority to engage in random stop-and-searches. Others, such as New York City's former police commissioner Howard Safir, urged the integration of facial imaging software into a system of video street surveillance so that pedestrians could be "compared with photographs of known terrorists" (Barstow 2001, 1:16), a technology already used in Britain (Rosen 2001) and at the 2001 Super Bowl, where petty thieves were picked out of the crowd of fans entering the stadium (*Los Angeles Times* 2001).

A graphic covering much of the page indicates in detail just what might be in store for New York's public space (and by extension those of other American cities): face-recognition cameras on lamp poles; police or security officers on every corner; dogs and their handlers roaming the squares and parks; reinforced, more bunker-like buildings; traffic restrictions sensitive to changing conditions (through the use of automatic barriers that can rise up through the pavement and close off streets nearly instantly); the elimination of "all above- and below-ground parking" near key public spaces and important buildings; continual broadcasts of public-service announcements throughout public squares (much like the messages that are broadcast in airports telling citizens to be on the lookout); and the installation of numerous planters, bollards, and blast-resistant trash cans. But interestingly, at the same time all of this is being proposed, Safir (one of the key architects of New York's "quality of life" policing campaign of the 1990s that sought to stringently police public spaces in the presumed interest of the safety of middle- and upper-class residents and visitors) argued against adopting a "bunker, bomb-camp mind-set" (quoted in Barstow 2001, 1:16).

This sentiment was echoed in the *Times* a week later by the Cooper Union's acting dean of architecture, Anthony Vidler (2001), but his inflection was decidedly different. Briefly reviewing the history of 20th century urban development and its relationship to ongoing fear of attacks (ranging from concerns over Zeppelin bombardments to IRA bombings),[1] a history that has brought with it a strong impulse toward metropolitan deconcentration, Vidler suggests that in the wake of the terrorist attacks "there will be an understandable impulse to flee" the city (Vidler 2001, 4:6).[2] But, Vidler argues, other cities' experiences with terrorism suggest that, in fact, "terrorism alone will not decrease the importance of city centers for the public life of societies," because "real community, as evident over the last week [of spontaneous public gatherings and memorials], is bred in cities more strongly than suburbs" (Vidler 2001, 4:6).

Vidler (2001 4:6) paints a decidedly different picture of public space than those "security experts" who see it as a threat: "The street as a site of interaction, encounter and the support of strangers for each other; the square as a place of gathering and vigil; the corner store as a communicator of information and interchange. These spaces, without romanticism or nostalgia, still define an urban culture, one that resists all effort to 'secure' it out of existence." Rather than wonder how public spaces can be made secure and how much it might cost (the experts in the earlier *Times* article estimated that "you could do a hell of a job with less than a billion dollars" [quoted in Barstow 2001, 1:16]), Vidler argues that true security—or at least an urban life worth living—consists in publicness itself. The sorts of proposals put forth by the panel of security experts, Vidler insists, would create "a world hardly worth living in and would inhibit the very contact through density that cities encourage" (Vidler 2001, 4:6). He goes on to argue that "urban public space has suffered major onslaughts in the last 20 years, from the increasing privatization encouraged by reliance on Internet services to the expansion of the mall-effect—whereby only the largest consumer outlets survive. In the current crisis, it is all the more important that the idea of public space, and its relations to urban community be sustained" (Vidler 2001, 4:6).

This book is about that "idea"—and, even more importantly, the *practice*—of public space in American cities. The terrorist attacks on September 11, 2001, did not so much launch a new debate about public space as serve to intensify one that already exists—and has existed for as long as there has been the "democratic" city. It is a debate—or more accurately an ongoing social struggle—that flares up, in varying forms, throughout the course of the 20th century (and in fact earlier too): as antiwar activists take to the streets; as labor activists seek to make the space necessary to press their claims; as free speech activists occupy ground meant for official pronouncements; as women make a space for themselves as part of "the public;" and as city after city tries to decide what to do about the homeless, about teenagers, and about other "undesirables." The question that drives this book is the question of who has the *right* to the city and its public spaces. How is that right determined—both in law and on the streets themselves? How is it policed, legitimized, or undermined? And how does that right—limited as it usually is, contested as it must be—give form to social justice (or its absence) in the city?

Much early commentary in the aftermath of the terrorist attacks suggested that Americans (and perhaps especially foreigners living in America) had better prepare themselves for the inevitable elimination of certain civil liberties. As the *Times* noted, even the American Civil Liberties Union was refraining from its usually automatic denunciations of such talk (Barstow, 2001). But what frequently gets lost in such discussions is the degree to which those liberties are *always* contested, always only proven in practice, never, that is, guaranteed in the abstract. Rights, as we will see in Chapter 1, do not work that way. For homeless people civil liberties and the right to public space have already been all but eliminated in the interests of enhancing the quality of urban life—and the "security"—for housed residents and visitors. For various movements of free speech—in the 1910s no less than the 1960s—the right to speak has often been undermined by spatial restrictions on *where* one can speak. For workers—and in recent years for anti-abortion picketers—the very act of picketing has frequently been declared by no less than the Supreme Court of the United States to be a violent act. If, as Anthony Vidler suggests, the idea of public space and its role in urban life needs to be preserved, then we also need to be aware that that idea has *never* been guaranteed. It has only been won through concerted struggle, and *then*, after the fact, guaranteed (to some extent) in law.

If Vidler's (2001, 4:6) vision of the city—especially his call to "search for design alternatives that retain the dense and vital mix of uses critical to urban life, rethinking the exclusions stemming from outdated zoning, real estate values and private ownership"—is attractive (and I think it is), then the need to continue to struggle over and for public space is now greater than ever. The automatic impulse during the current sense of emergency is to defer to the security experts and their vision of the orderly and safe city. The alternative implicit in Vidler's vision seems, by comparison, highly unpalatable. The sort of city he promotes must necessarily retain some tolerance for risk and danger. It must take for granted that at least some level of "fear" will always be present in urban life. There is no way around that, as unattractive as such a vision had become even before September 11. Struggle—which is the only way that the right to public space can be maintained and the only way that social justice can be advanced—is never without danger of violence. How that potential for violence is policed, encapsulated in law, sublimated in design, or turned toward either regressive or progressive ends makes all the difference in the world.

My goal in this book is to examine some of the contours of that struggle over the past hundred years in American cities. Much of what follows has been previously published, though all is rewritten and updated, sometimes extensively, sometimes only a bit. My reason for bringing this work together in a single volume is to make an argument about the tenuous nature of what the French Marxist and social theorist Henri Lefebvre called "the right to the city." That right, as I hope becomes clear in the course of this book, is dependent upon public space. But just what public space is—and who has the right to it—is rarely clear, and certainly cannot be established in the abstract. I present in rough historical order, therefore, a series of linked case studies that explore the relationship between social exclusion, social rights, and social justice in American public space. The links between these studies are many, but include a concern with the relationship between social activism and changes in public space law; the role of marginalized actors (migratory workers, the homeless) as a focus of social exclusion; the need not just to produce public space (as so much work in geography has studied) but to actively *take* it, if a claim of right is to be made; and a set of questions about the dialectic of order and *dis*order as it is worked out in specific places at critical times—the streets of San Diego in 1912 as the Industrial Workers of the World were on the march, People's Park in Berkeley, California, in 1969 as an imagined alternative to alienated bureaucratic society, the Civic Center of Santa Ana in the 1990s when the homeless were seen as only so many broken windows marring a landscape poised for economic revitalization, or New York City in those booming years immediately before September 11 when the "quality of life" was up for grabs.

Chapter 1 lays out a theory of social justice as it relates to urban public space. Working through a striking argument by Raymond Williams about Matthew Arnold's reactions to the Hyde Park riots of 1866, I examine some recent statements by commentators that Williams would have called "little Arnolds"—those who see order as properly trumping rights in urban space in nearly all instances—to show how debates and struggles over who has access to public space, and who is excluded, define the "right to the city"—and why a right to the city must be at the heart of any vision of a progressive, democratic, and just world. Part of my argument is that, in fact, rights matter (a position with which some on the left disagree)—and so does law. I suggest that "rights talk"—and even more the practical assertion of rights—remains a critical exercise if social justice is to be advanced rather than constricted.[3] Such a claim demands at least a brief indication of how social justice should be theorized, and so the chapter concludes with a discussion of the relationship between rights, social justice, and urban space.

Social justice, rights, and their relationship to urban space, as I have indicated, are not determined in the abstract, but rather in practice. So, in Chapter 2 I turn to these practices. In particular, I try to show how specific social struggles over public space (and the assertion of rights therein) lead to transformations of public space law as courts seek to either adjudicate or eliminate conflict.[4] Sometimes the most important practices are not ones that we like. The most significant recent U.S. Supreme Court cases about protests in public space have concerned anti-abortion protesters outside clinics and the homes of abortion providers. These cases draw on a long history of case law that in fact has more to do with controlling (and sometimes eliminating) labor dissent than it does with the sorts of political dissent exercised by anti-abortion protesters. I trace this history from its origins in struggles over the right to speak on the streets by the Industrial Workers of the World around 1910, through a series of celebrated, and rather reactionary, cases concerning "subversive" protesters during World War I and labor picketers in the 1920s and 1930s, and to the eventual codification of what came to be known as "public forum doctrine" in the 1950s and 1960s. These struggles and court cases all involved the right to speak— or what we so often carelessly call free speech—in public spaces, and they involved a rather torturous, but still exceedingly important, distinction the Supreme Court has made between "pure speech," "expressive conduct," and behavior. This distinction is important because it helps *limit* rather than open up what can be said and done in public space and thereby helps to protect the interests of capital and the state.

By 1939, the U.S. Supreme Court had codified, and presumably vouchsafed, the right to speak in public spaces. But that did not stop (and has not stopped) innumerable jurisdictions from hemming in that right, and often eliminating it altogether. Sometimes the means of elimination is expressly geographical. Jurisdictions often try to "protect" the right to speech in public space by assuring that it occurs in such out-of-the-way places that it has little chance of being heard. Such was the case at the University of California at Berkeley in the early 1960s. Conflict over the right to speak became a conflict over who controlled specific spaces (and indeed over the content of that control). Chapter 3 examines this episode and some of its consequences.[5] The Berkeley Free Speech Movement did not inaugurate campus activism in the 1960s (its roots lie more in the civil rights movement), but it did solidify certain aspects of it, including the implementation of a critique not only of politics and justice in America (and beyond), but a critique of modern alienation—a critique that blossomed most fully, perhaps, in the streets of Paris and Prague during the spring of 1968 (and which Lefebvre was anticipating with his polemic on the right to the city). In Berkeley, this critique eventually coalesced into the Peoples Park movement, and so the chapter ends by exploring the roots of and early struggles over People's Park.

Activists established People's Park (and baptized it in riot) as what they hoped would be a small unalienated space within a city as a whole defined by alienation. As such, it became a refuge for many of the outcasts of society, including the homeless. By the 1980s, according to many (including some nearby residents, merchants, and the university), the sheer number of homeless people in the park had allowed it to

degenerate from a space of freedom to a space of depravity. As a consequence, the university and the city of Berkeley entered into an agreement to redevelop the park and thereby to discourage its "inappropriate" use, especially by homeless people. I argue in Chapter 4 that this
plan, and the riot that it touched off in 1991, illustrate just the sort of
dichotomy in our ideas about public space noted above in the contrast
between the vision of public space advanced by "security experts" and
that advanced by Anthony Vidler. This chapter is an updated and revised version of an article published in 1995 (Mitchell 1995).[6] I used
the article, and now the chapter, to raise questions about the *kinds* of
public spaces available in the contemporary world, and thus the limits
on what sorts of "publics" can be formed. I assess the argument that we
have reached the "end of public space" in the contemporary city and
find that that argument is overly simplistic, since it assumes that public
space already simply existed (rather than was socially produced through
struggle). It is also overly simplistic in that it does not necessarily appreciate how new kinds of spaces have developed, creating new opportunities for publicity. One of those kinds of public space, of course, is
the space of the electronic media. What I found most interesting as I
revised my argument for this volume was the degree to which I underestimated the privatization of, and economic control over, the World
Wide Web, and hence overestimated its potential as a "public space"
and the sorts of democracy that public space helps advance. And yet, at
the same time, I also underestimated electronic media's role in *organizing* political action, and hence the possibility of democratic struggle in
urban public spaces. I try to address this dual underestimation in the
present version.

But my real interest lies with the fate of homeless people in material, urban public spaces and the ways that our efforts to eliminate them
from those spaces are leading to a remarkably constricted public sphere
and a rather shriveled notion of rights, a notion that radical workers in
the first half of the century—like those for whom much public forum
doctrine was written—would not much recognize (but also would not
be much surprised by). Chapters 5 and 6 turn to these issues of homelessness, public space, and rights in the contemporary city. Chapter 5
returns us, also, to issues of law and its role in affecting social and political exclusion by examining the roots and consequences of the implementation of anti-homeless laws around the country.[7] In this chapter I
build on the preceding one to explore how movements to regulate pub-

lic space so as to remove homeless people lead to a specific, and highly constricted, sort of public sphere. I do not find this kind of public sphere to be very attractive. Indeed, to me it speaks of a highly sanitized city and a fully deracinated politics—a politics that elevates the importance of aesthetics over the needs of some people simply to survive. My main point in this chapter is that the new spate of anti-homeless laws and other "quality of life" initiatives in the contemporary city rely on *fear* as a driving force and thus tend toward not only the sort of security state that the New York experts outline but also the wholesale elimination of a class of people who have nowhere else to *be* but in public. In short, anti-homeless laws undermine the very right to the city.

Chapter 6 continues this analysis but places anti-homeless laws within two contexts: the "broken windows" theory of policing and urban policy that has provided their justification; and a recent proposal to begin zoning public spaces.[8] I argue that the current city is one in which the upper hand of "justice" itself has been firmly taken by the urban right, forcing the left into arguing, at best, for some awfully paltry rights—such as the right to sleep on the sidewalk, the right to beg, and the right to urinate in an alleyway. These are hard rights to either get excited about winning, or in fact to continue struggling for, no matter how absolutely central they may be in the current political-economic climate where the right to housing, to a livelihood, or to decent physical and mental health care has simply been wiped from the agenda by the stunning success of neoliberalism. My focus on homeless people in public space—and the attempts to regulate them out of existence—in these chapters derives from the fact that *homelessness* has become so prototypically the bellwether of urban justice. If anyone needs the right to the city, surely it is the homeless. But such a right must entail not just the right to sleep or piss in public but also the right to *inhabit*, to appropriate, and to control. And it must be affected through a radical wresting of power and a much fuller *democratization* of public space. Neoliberal urban reform calls for the constant increase of urban order. Struggle for social justice in the city—for the right to the city—must therefore seek to establish a different kind of order, one built not on the fears of the bourgeoisie but on the needs of the poorest and most marginalized residents.

By way of conclusion, therefore, I turn to a brief examination of how that order should be conceptualized. I do not suggest what it should be, for any social order will be a product of social practice—the politics of the street. It is not simply the result of normative argument, even though any politics of the street, as I hope this volume makes clear, is always mediated through normative argument. And so, in the Conclusion, I do suggest that certain *forms* of normative conceptualization of the city and of public space—indeed, certain utopian images of what the city could or should be—have been and remain crucial in these politics of the street. While much of my analysis in this volume may seem pessimistic, the undercurrent of each of the chapters is that social action—protest, the bringing of lawsuits and other legal actions, the active *taking* of space—has been the fulcrum upon which the right to the city has been leveraged, both in its actual (limited) practice and in the way it can serve as a beacon for a more open, more just, more egalitarian society. The undercurrent of radical activism that shapes space in and against the "regimes of justice" that regulate it should not be underestimated or dismissed, no matter how often such activism is either defeated or co-opted. Where I see hope is in exactly those moments when radical activist movements have arisen—again and again—to take back the city and to make into something better, movements that "rethink the exclusions" of the past (as Vidler puts it)—and that struggle to *remake* the city in a more open and progressive light.

Remaking the city in an image of openness and justice in the wake of September 11 will be harder than ever. But it is also more necessary than ever. One of the common refrains in those shocked days immediately following the attacks was that "everything has changed." Yet, as Bertell Ollman (1990) never tires of reminding us, change is the norm; what needs to be explained is fixity. If that is the case, then public space, solid as it is (and its materiality *does* matter immensely), is ever in a state of flux, ever subject to reformation. What September 11 has likely done is deepen tendencies already in place. Those arguing for security and order over openness and the messy risks of street politics will find further reasons and means for advancing their agenda. At the same time, the prodemocracy, anticapitalist movement has been sent scrambling, wondering not only what its object is but also whether it can ever protest in the streets again. But that too is just a deepening of trends, as the increasingly violent response to protesters in Washington, Quebec, and Genoa during the first part of 2001 made clear. Finally, there are the trends in the content and shape of public space itself. We were already moving toward the sorts of mall-like public spaces that Vidler notes in his *Times* article, toward a sort of suburbanization of downtown. Even the largest of new public spaces, like the soon-to-be-built Downsview Park in Toronto, are more and more modeled not on an ethic of interaction but an ethic of seamless, individuated movement and circulation: public interaction based on the model of commodity and capital flows.[9]

I make no sure guesses as to what the future holds. But I do have some sense of—and I hope this volume helps to convey—what the past, and past struggles over public space, have held. If any of the events, trends, and struggles I have identified in this book have value, my hope is that it will be in pointing us both to the degree to which public space is always an achievement (invariably against very steep odds) and *therefore* to what a just city could be. Lefebvre argues that the right to the city is "like a cry and a demand." Now, more than ever, that cry, that demand, must be heard. And it must be put into practice.

NOTES

1. Interestingly, Vidler does not discuss the decentralization schemes sponsored by the U.S. government out of fear of nuclear attack during the cold war. Such schemes—ranging from the development of the interstate system for defense purposes to the planned deconcentration of industry in Detroit—have had a profound effect on American cities.
2. Certainly aware that the prominent "new urbanist" James Howard Kunstler (writing with Nikos Salingaros) had already done so in a widely circulated essay written within a day of the attack, Vidler suggested that "the 'new urbanism' movement, with its low-density developments like Seaside or Celebration in Florida, designed to replicate small-town life in pre-modern America, will no doubt take the opportunity to denounce tall buildings as inherently mistaken." A version of the Kunstler and Salingaros essay, "The End of Tall Buildings," and some responses to it, can be read at *http:// www.peoplesgeography.org/space.htm*.
3. The argument I make about rights in Chapter 1 was first laid out in Mitchell (1997a); the remainder of the chapter is new.
4. Chapter 2 is a significantly revised and updated version of Mitchell (1996b).
5. This chapter is a revised version of Mitchell (1992). In particular, I fill in and update some of the historical-geographical detail of just what happened on the Berkeley campus in the fall of 1964.
6. Among other things, I update the current status of the park. Once again, UC is considering building dormitories on the land occupied by the park, which was precisely the plan that set off the 1969 riots in the first place.
7. Chapter 5 is a slightly revised version of Mitchell (1997b).
8. While most of the analysis in this chapter is new, the examination of "bro-

ken windows" and public space zoning revise an argument I put forth in a recent book chapter (Mitchell 2001b). That chapter asserted that we now live in a "post-justice" (rather than the more neutral "postmodern") city, an argument I do not develop in the present context.

9. Richard Van Deusen and I evaluate the plans for Downsview in a recent chapter (Mitchell and Van Deusen 2002) of a book (Czerniak 2002) that explores the design competition for the park. The competition itself, and the sorts of plans it called forth, is deeply indicative of current corporate public space ideology.

To Go Again to Hyde Park
Public Space, Rights, and Social Justice

Public space engenders fears, fears that derive from the sense of public space as uncontrolled space, as a space in which civilization is exceptionally fragile. The panic over "wilding" in New York City's Central Park in the late 1980s (rampaging young men violently terrorizing joggers and other park users for the sheer joy of it), the fright made palpable by the explosions in Atlanta's Olympic Park in 1996, and the newfound fear of public space spurred by the sense of vulnerability attendant upon the September 11, 2001, terrorist attacks, no less than the everyday gnawing uneasiness we feel when we step around a passed-out homeless person on a sidewalk, often convince us that public space is the space of anarchy. Such an association of public space with anarchy is, of course, not new; it is not just a feature of the contemporary city, of the current media-encouraged, overweening concern about crime, homelessness, and random terrorism that makes public space seem such an undesirable attribute of the contemporary American city.

Raymond Williams (1997 [1980], 3–5) reminds us, for example, that Matthew Arnold's (1993) famous declaration in *Culture and Anarchy*—that culture represents (or ought to represent) "the *best* knowledge and thought of the time" (1993, 79)—was made in response to working people forcing their way into Hyde Park in 1866 to hold an assembly in support of the right to vote. For Arnold, the Hyde Park demonstrators were "a symptom of the general anarchy" (Williams 1997

13

[1980], 6) rather than people struggling for their rights—their right to assemble, their right to speak, their right to vote. A Hyde Park "rioter," according to Arnold, "is just asserting his personal liberty a little, going where he likes, assembling where he likes, bawling where he likes, hustling as he likes" (Arnold 1993, 88, quoted in Williams 1997 [1980], 6).[1] Even more—and even more shrilly—Arnold objected to a working person's "right to march where he likes, meet where he likes, enter where he likes, hoot as he likes, threaten as he likes, smash as he likes. All this, I say, tends to anarchy" (Arnold 1993, 85, quoted in Williams 1997 [1980], 6).

The proper response, according to Arnold, was repression, the reigning in of "rights," and the asserting of firmer control over public space, for "without order there can be no society; and without society there can be no human perfection" (Arnold 1993, 181, quoted in Williams 1997 [1980], 6). Only with order can culture flourish, can cities be centers of civilization.[2] Williams's point in resurrecting the context of Arnold's arguments about culture is important: those rights we take as "immemorial," such as the right to assemble in and use public space, are not only relatively new, they are always hotly contested and only grudgingly given by those in power. *Always* hotly contested: rights over and to public space are never guaranteed once and for all. New struggles emerge, if not only over the right to vote then over the right to live a sane and peaceful life in the nuclear age, the right to control over government in totalitarian states, or, especially in the "postmodern" cities of the Western world, the right, in the absence of decent, affordable housing, simply to *live*.[3] As Williams (1997 [1980], 8) rightly proclaims: "it will always be necessary to go again to Hyde Park."

But, just as it is always necessary "to go again to Hyde Park"—for people to take control of public space in defiance of the order, control, and contempt imposed upon them in the name of vouchsafing the vested interests of the few—so too in response do there arise legions of Matthew Arnold imitators. "Our own little Arnolds," Williams (1997 [1980], 8) called them, who claim they are promoting "excellence and humane values on the one hand; discipline and where necessary repression on the other." It is not just spectacular protests, riots, or mass demonstrations that draw out these "little Arnolds." In the contemporary United States, these "little Arnolds" have multiplied most rapidly around the perceived disordering of city streets that has come with the persistent growth of homelessness, with the growing numbers of the un- and underemployed, the mentally ill, and the drug-addicted who have no other recourse than to live their lives in full view of the urban public. For the homeless "to go to Hyde Park" is often a matter of survival; for their detractors this "occupation" of public space by homeless people is seen as a clear affront to the order, dignity, and the civilization of the city.

In the United States, where the crisis of homelessness is now beginning its third decade, perhaps best known among these "little Arnolds" is the nationally syndicated newspaper columnist and regular television talk-show guest George Will. Will frequently uses his newspaper column to promote the idea that the need to maintain "order" and "civility" in public space is simply commonsense. Those who work to promote the rights of homeless people to use public space (as a refuge, as a place to sleep, as a stopping point, as a place of community and conviviality) are nothing more than "gladiators of liberation"[4] engaged in the "business" of "abstract compassion" (Will 1995, 7B). Over the course of the homelessness crisis, Will has been impressively consistent. For nearly two decades, bidding fair to be our era's Matthew Arnold, our era's defender of the sweetness and light that is his version of culture and civility, he has argued forcefully that the need for "order" trumps individual or collective liberty (see, e.g., Will 1987, 1995, 1997). Will, echoing the comments of Anatole France, but with none of the latter's piercing irony, is fond of asserting that there simply "can be no right to sleep on the streets" (Will 1987). The need for a certain kind of collective order outweighs whatever putative right a homeless person might have to find a space for living in the public spaces of the city. The right for the housed residents and visitors of a city to move about without encountering any sights that might trouble them outweighs the right of a homeless person to urinate in a park or alley when there are no public toilets and she or he has been barred admission to restaurants or other semipublic places. The need for order, the need to guard against anarchy, demands at least that much, according to George Will.

Will is hardly alone in his arguments. Rather, he is supported by the concentrated energies of such organizations as the Manhattan Institute (a conservative think tank in New York),[5] the American Alliance for Rights and Responsibilities (a conservative public interest lawyers guild), nationally based policy and opinion mills like the Heritage Foundation and the American Enterprise Institute, and big-city mayors from New York and Cleveland to Los Angeles, San Francisco, and Seattle. When Will argues that there can be no right to sleep on the streets, Robert Tier (1993, 287) of the American Alliance for Rights and Responsibilities[6] echoes him (and Matthew Arnold) by pointing out that what is at stake is actually not rights at all, but a question of choices. He argues that while the struggle for civil rights might once have had a place in American society (it "helped end American apartheid"), hard-won civil rights are now "used to try to trump many legitimate community interests, and to elevate all kinds of individual *desires* into assertions of rights. They are now used to defend the colonization of parks by people *wishing* to sleep there, to assert the right to sleep and eat in the public place of one's *choosing*, and to beg in any way one *pleases*" (emphasis added). The problem of homelessness, according to Tier, is not a lack of affordable housing or decent public services, but one of "civility." Adopting a language of inclusiveness, Tier (1998, 290) argues that laws restricting homeless people open public space up for *all* to enjoy: "those with Armani suits and those with nose rings; elderly people and gay couples; residents and visitors; rich, middle, and struggling classes" (but presumably not those who have no other place to *be* but in the public spaces of the city). The means of assuring such an open and accessible space, such a civil space, Tier (1998, 290) continues, is to practice "tough love." Rather than working toward the construction of a vibrant public housing program that would make housing affordable; rather than fashioning a decent mental health system that would make the "care" that Tier advocates better than the disease (see Winerip 1999); rather than seeking ways to transform an economic system that requires high levels of structural unemployment to function, Tier (1993, 291) argues that we need "the protection of pedestrians from unwanted solicitations, harassments, and assault." Government should promote the interests of some, Tier is suggesting, even if doing so requires undermining the even more basic rights of others. Tier's (1993, 286) "call for public order" to counteract the descent of urban public space into anarchy shades quickly into repression. No further evidence is needed than New York Mayor Rudy Giuliani's (in office from 1993 to 2001) order to the police, in November 1999, to arrest any homeless or other street people who did not "move along" when told to do so, even if they committed no crime (a practice a federal appellate court in New York had several years earlier already declared unconstitutional).[7]

The desire to control the streets and other public spaces of the city is not limited to issues concerning homeless people. "Our own little Arnolds" have wider targets. As Mayor Giuliani has made clear, the desire to counteract "anarchy" with repression runs the gamut of public space uses from rallies and demonstrations (as with the police department's violent response to both the "Million Youth March" and the Matthew Shepard memorial march in 1998),[8] to ridding the streets of unlicensed peddlers, to a crackdown on public "vice" throughout Manhattan, to destroying community gardens so as to hand over the property they occupy into the waiting arms of private developers.[9] Often this assault on homeless people, community gardeners, small-time peddlers, and young people seeking a place to hang out is couched in the language of liberty.[10] Without order, the argument (from Arnold, through Will and Tier, to Giuliani) goes, liberty is simply impossible. And that order must be explicitly geographic: it centers on the control of the streets and the question of just *who* has *the right to the city*.

PUBLIC SPACE AND THE RIGHT TO THE CITY

"The right to the city" is a slogan closely associated with the French Marxist philosopher Henri Lefebvre. Writing on the 100th anniversary of the publication of the first volume of *Capital* and just before the student and worker uprising of May 1968, Lefebvre's short book, *Le droit à la ville*, sought to outline what a specifically urban postbourgeois philosophy might be. Much of the book (now published in English as part of a collection of *Writings on Cities*: Lefebvre 1996) is highly abstract and arcane, little more than a set of notes, many of which would later be expanded upon in Lefebvre's (1991 [1974]) magnum opus, *The Production of Space*.[11] But within this rather arch argument about the content of philosophy and its relationship to the changing social relations of cities were a set of aphorisms and a key set of concepts that had immediate popular resonance. The most important is Lefebvre's normative argument that the city is an *ouvre*—a work in which all its citizens participate.

There are several issues here that are critical to the development of the argument about public space and social justice that I will make in this book. The first is Lefebvre's insistence on a right to the *city*. Lefebvre was deeply attached to the rural countryside, especially the village of his birth (Merrifield 2002; Shields 1998), but he shared with

Marx a disdain for the *idiocy* of rural life. Idiocy in this sense does not refer to the intelligence of the inhabitants, or even the nature of their customs, but to the essential *privacy*—and therefore isolation and homogeneity—of rural life. In contrast, cities were necessarily *public*—and therefore places of social interaction and exchange with people who were necessarily different. Publicity demands heterogeneity and the space of the city—with its density and its constant attraction of new immigrants—assured a thick fabric of heterogeneity, one in which encounters with difference were guaranteed. But for the encounter with difference to really succeed, then, as we will see in a moment, the right to *inhabit* the city—by different people and different groups—had always to be struggled for. This is the second issue. The city is the place where difference *lives*. And finally, in the city, different people with different projects must necessarily struggle with one another over the shape of the city, the terms of access to the public realm, and even the rights of citizenship. Out of this struggle the city as a work—as an *ouvre*, as a collective if not singular project—emerges, and new modes of living, new modes of inhabiting, are invented.[12]

But the problem with the bourgeois city, the city in which we really live, of course, is that this *ouvre* is alienated, and so not so much a site of participation as one of expropriation by a dominant class (and set of economic interests) that is not really interested in making the city a site for the cohabitation of differences. More and more the spaces of the modern city are being produced *for* us rather than *by* us. People, Lefebvre argued, have a right to more; they have the right to the *ouvre*. Moreover, this right is related to objective needs, needs that any city should be structured toward meeting: "the need for creative activity, for the *ouvre* (not only of products and consumable material goods), the need for information, symbolism, the imaginary and play" (Lefebvre 1996 [1968], 147). More sharply: "The right to the city manifests itself as a superior form of rights: right to freedom, to individualization in socialization, to habitat and to inhabit. The right to the *ouvre*, to participation and *appropriation* (clearly distinct from the right to property), are implied in the right to the city" (Lefebvre 1996 [1968], 174).

When it was published, this call—this cry and demand—for the right to the city resonated immediately, because Lefebvre was clearly reflecting, and reflecting on, the growing season of unrest that was the 1960s. From his (often uneasy) links to the Situationist International and other radical groups in Paris (see Jappé 1999; Merrifield 2002;

Shields 1998), Lefebvre took to heart the argument that current *situations*—current geographical spaces—had to be radically transformed if the project of human emancipation was to be advanced. As the Situationist Guy Debord (1994 [1967], 126) was simultaneously arguing, "the proletarian revolution is that *critique of human geography* whereby individuals and communities must construct places and events commensurate with the appropriation not just of their labor, but of their total history" (emphasis in original).

This was not just philosophical positioning or radical posturing by either Lefebvre or Debord. From the civil rights movement, the Port Huron Statement of the Students for a Democratic Society, and the Berkeley Free Speech Movement in the United States, to the stirrings of the anti-war and anti-imperialism movements that were in fact global in reach, to the specific complaints of Parisian students fed up with being molded into uncomplaining "organizational men" (and women), radical social transformation really seemed possible. And for Lefebvre, this implied the development (finally) of a fully urban society. The right to the city was the right "to urban life, to renewed centrality, to places of encounter and exchange, to life rhythms and time uses, enabling the full and complete *usage* of . . . moments and places . . . " (Lefebvre 1996 [1968], 179). That is to say, the *use-value* that is the necessary bedrock of urban life would finally be wrenched free from its domination by exchange-value. The right to the city implies the right to the uses of city spaces, the right to *inhabit*. In turn, and highly germane to the current American city, where we are reduced to arguing over whether one has the right to publicly urinate if he or she is homeless (Mitchell 1998a, 1998b), the right to inhabit implies a *right* to housing (Lefebvre 1996 [1968], 179): a place to sleep, a place to urinate and defecate without asking someone else's permission, a place to relax, a place from which to venture forth. Simply guaranteeing the right to housing may not be sufficient to guaranteeing a right to the city, but it is a necessary step toward guaranteeing that right.

That is to say, the right to housing is one form of *appropriation* of the city, and that is why Lefebvre was at pains to set this off from the right to property. For property, of course, is the embodiment of alienation, an embodied alienation backed up by violence (Blomley 1998, 2000a; 2004; Rose 1994). More accurately, property *rights are* necessarily exclusive: the possession of a property right allows its possessor to exclude unwanted people from access (Blomley 2000b, 651;

MacPherson 1978). And this act of expulsion, this right of property, Blomley (2000a, 88) notes, frequently involves invoking the power of the state: "Police can be called to physically remove a trespasser; injunctions prepared, criminal sanctions sought. As such, expulsion is a violent act. Violence can be explicitly deployed or (more usually) implied. But such violence has state sanction and is thus legitimate." This issue is particularly important in a world where some members of society are not covered by *any* property right (Waldron 1991) and so must find a way to inhabit the city *despite* the exclusivity of property—either that, or they must find ways, as with squatting, and with the collective movements of the landless, to undermine the power of property and its state sanction, to otherwise appropriate and inhabit the city. In the contemporary city of homelessness the right to inhabit the city must always be asserted not within, but against, the rights of property. The right to housing needs to be dissociated from the right to property and returned to the right to inhabit.

In the United States and Canada, Lefebvre's concerns about the right to *inhabit* the city were echoed by the radical geographer William Bunge in his "expeditions" in Detroit and Toronto. Bunge argued strongly for the rights of resident communities over "foreign invaders" (such as outside capital, suburban commuters, etc.) and for the rights of people over machines (such as cars) (Bunge 1971; Bunge and Bordessa 1975; Horvath 1971, 1974; see Merrifield 1995). Radicalized by the Detroit riots of 1967, and working both among and for community activists, Bunge focused on the urban life spaces of poor children, particularly African American children. His *Fitzgerald: The Geography of a Revolution* (1971) is certainly a cry and a demand—and a brilliant exploration into the daily geography of life in a disinvested, redlined, violent capitalist city. The right to *inhabit* was tenuous, and Bunge exposed its nature in the American city through a remarkable cartography of rat bites, broken glass, empty lots, and hit-and-run accidents. The right to the city, the right to inhabit—such as it was—existed within a web of violence and deprivation, a web of violence and deprivation that, as Blomley (2000a) makes clear, is both in part a result of, and also hidden by, the seeming naturalness of property.[13]

Central to Bunge's revolutionary geography, as to Lefebvre's, was the right to housing and the right to control over public space. In the first case, for Bunge and Bordessa (1974), housing was not only desperately needed but also desperately in need of redesign. Public housing in Detroit and Toronto and other North American cities was designed so as to be "anti-children." High-rises, for example, had no space for play—at least no safe and decent space. Likewise what sufficed as public space in the inner city—empty lots, busy streets, and barren, windswept, unsupervised playgrounds—represented more a geography of death than of life. The right to housing, the right to inhabit the city, thus demands more than just houses and apartments: it demands the redevelopment of the city in a manner responsive to the needs, desires, and pleasures of its inhabitants, especially its oppressed inhabitants. Against this, the actually existing city, the arguments of the "little Arnolds," with their overweaning concern for order and "civilization," with their defenses of the geography of privilege, begin to seem paltry and mean-spirited at best. The cry and the demand for the *right* to the city is the best means there is to begin to assure what Bunge has called "the geography of survival."

"Rights Talk"

If such arguments about *rights* were of great importance at the time of the upheavals of 1968 and if they remained important to social movements and activists in the 1970—if, that is, the discourse about rights was central to those particular "returns to Hyde Park"; then such arguments, such discourses, have in more recent years fallen out of favor among much of the academic left and to some degree urban social movements, as well. Complaints about the limits of "rights talk" to progressive social change are legion and are clearly linked to the rise of a more "postmodern" discourse in the wake of the defeat of the 1968 uprisings. The defeat of the left after 1968—perhaps a little later in the American context, even though the relatively conservative Richard Nixon was elected president that same year—indicated to many that the left had for too long hung its hopes on unrealistic (and ontologically suspect) universalist notions of social justice and emancipation. The "enlightenment project," subscribed to by everyone from the signers of the Declarations of Independence, the Declaration of the Rights of Man, and the Declaration of Sentiments to Karl Marx and Martin Luther King, had proven itself to be, many argued, not just easily corruptible by, but actively complicit in, the rise of fascism, the development of weapons of mass destruction, and the transformation of people not into enlightened subjects but passive bearers of techno-bureaucratic rationality.

Often labeled "postmodernist," especially after Lyotard's (1985) broadside against "metanarratives,"[14] such skepticism toward "rights talk," like the call for repressive order in public space, is not in fact particularly new. Indeed, Marx himself was famously skeptical toward the value of "rights" as an organizing principle of social struggle. After all, when rights conflict (as they inevitably do) "force decides" (Marx 1987 [1867], 225). But, as David Harvey (1996, 345) correctly notes in regard to this passage, Marx's point was not at all to abjure completely the efficacy of rights (see also Harvey 2000). Rather, his point was that rights remain efficacious only to the degree they are backed by power, by at least the implicit threat of violence—violence that is at times the "property" of the state and at other times, and crucially, "extra-legal" (Harvey 1996, 346, following Derrida 1992, 35).[15] To put that another way, rights are at once a means of organizing power, a means of contesting power, and a means of adjudicating power, and these three roles frequently conflict. The difference between Marx's skepticism toward rights (and justice more generally: see Merrifield and Swyngedouw 1996, 1–2) and more postmodern skepticism of rights as a universalizing or totalizing discourse (Lyotard 1985) is that, while the latter sees rights' *indeterminacy* as their Achilles' heel, more Marxian (and hence more modernist) approaches are concerned with the degree to which rights, despite whatever degree of indeterminacy they may possess, are still to some degree *determinant* in social life. "Rights"—to the degree they are institutionalized and protected within specific social situations, to the degree that they are and are not backed by the violence and the power of the state, and to the degree that they protect the interests of some at the expense of others (despite and because of the universalizing qualities)—are social relations and hence a means of organizing the actual social content of justice.[16]

Yet, the specific arguments against rights as a focus of progressive social justice and political organizing are important and worth considering. The most cogent arguments against "rights" as a rallying cry for progressives are expressed in Mark Tushnet's (1984) landmark "An Essay on Rights."[17] A leftist law scholar, Tushnet argues that "rights talk" is merely distracting, turning progressive attention away from what really needs to be done in the interests of social justice: "People need food and shelter right now, and demanding those needs be satisfied— whether or not satisfying them can today be persuasively characterized as enforcing a right—strikes me as more likely to succeed than claiming that rights to food and shelter must be enforced" (Tushnet 1984, 1394). This is the case because rights suffer from four flaws that are, according to Tushnet, fatal. First, rights suffer from *instability*. That is, they are not universal and abstract, as the discourse about them often claims, but rather exist only as products of particular political and social moments. As these moments change, so too do the content of rights. Second, rights suffer from *indeterminacy*. That is, "the language of rights is so open and indeterminate that opposing parties can use the same language to express their positions" (1371). Third, rights suffer from *reification*. That is, they treat real, complex experiences as an instance of the simple exercise of abstract rights, which "mischaracterizes" (1382) and devalues those experiences, eliminating what is most important from any social action—its political efficacy. Finally, and most importantly, rights suffer from *political disutility*. That is, rights often protect privilege and domination instead of the oppressed and minorities, as when commercial speech or the ability of rich donors to buy candidates in an election is "guaranteed" by the First Amendment right to free speech.

Tushnet (1984, 1386) is forceful on this last point: "It is not just that rights-talk does not do much good. In the contemporary United States, it is positively harmful." This is so, in part, because

> The contemporary rhetoric of rights speaks primarily to negative ones.[18] By abstracting from real experiences and reifying the idea of rights, it creates a sphere of autonomy stripped from any social context and counterposes to it a sphere of social life stripped of any content. (Tushnet 1984, 1392–1393)

And equally important, "the predominance of negative rights creates an ideological barrier to the extension of positive rights in our culture." Yet, this is not just a matter of progressives needing to do a better job of promoting positive rights, for, as already noted, Tushnet argues that focusing political action on such a goal would detract from the struggle for the *real* needs at hand (such as food and shelter for the hungry and homeless).

In the context of the radical restructuring of both capital and the state that has quickened since the structural crises that came quickly on the heels of the 1968 uprisings, arguments such as those by Tushnet need careful attention. Such arguments raise precisely the sorts of questions that ought to be addressed by those who would fight for a more just world in the context of a reordered political, economic, and social life. In an argument against rights, targeted at a popular audience and anticipating much of the transformed discourse about class and economic power that crystallized in the anti-World Trade Organization (WTO) protests in Seattle at the end of 1999, Richard Rorty (1996, 15) asserts that the most serious problems facing the United States (and the rest of the world) center on the "power of the rich over the poor" and that "[a]s Karl Marx pointed out, the history of the modern age is the history of class warfare, and in America today, it is a war that the rich are winning, the poor are losing, and the left, for the most part, is standing by." The left is standing by, Rorty argues, because it has brandished a particularly weak weapon against the rich: rights. Rorty argues that "rights" are weak because they do not directly attack "economic injustice." Instead, they are, at least in the contemporary United States, concerned mostly with issues of cultural domination.[19] Therefore, Rorty suggests that both economic and cultural injustice is better attacked with the "robust, practical, and concrete language" of "moral discourse" (Rorty, 1996, 16).[20]

Like Rorty, Tushnet (1984, 1402) hangs his argument on the (hard to dispute) claim that "[t]hings on the whole are terrible" since the United States (or, more accurately, capital coordinated through the political and military might of the United States) has created "one of the great empires in world history [where] life in the metropolis goes on as well as it does only because the metropolis exploits the provinces." Under such conditions (conditions that we now have grown accustomed to signifying with the deceptively neutral name "globalization"), "rights talk" seems decidedly secondary. There are far more important and immediate battles to be fought: the battle for housing, for income redistribution, for worker power, against corporate colonization of every aspect of our lives, against racism, against sexism, and against the homophobia that rules everyday life.

Yet, for all the strength of Tushnet's (and Rorty's) analysis, something is lacking—something that the current restructuring of capital and the state makes so readily apparent that it is hard to see how such perceptive analysts can so easily sidestep it. And that is simply this: at a time when the globalization of capital is aided and abetted every step of the way by what Stuart Hall (1988) famously called (in the British case) "authoritarian populism";[21] when, under the name of free trade and unfettered markets, capital is free to systematically crush any vestige of social life not yet under its sway, free to create a world in which the immiserization of the many so as to enrich the very few is packaged as inherently *just* (and liberatory); then those who seek to create a better world have few more powerful tools than precisely the language of rights, no matter how imperfect that language may be (Blomley 1994b).[22] Rights establish an important *ideal* against which the behavior of the state, capital, and other powerful actors must be measured—and held accountable. They provide an institutionalized framework, no matter how incomplete, within which the goals of social struggle can not only be organized but also attained. As Iris Marion Young (1990, 25) argues, "rights are relationships not things; they are institutionally defined rules specifying what people can do in relation to one another. Rights refer to doing more than to having, to social relations that enable or constrain action."

To put that another way, and to generalize the point, there is a central contradiction that all social movements must face, a contradiction that must be faced squarely even though it is hard to see just how it can be overcome. On the one hand, one of the greatest impediments to freedom, to a just social life, to the kind of world Rorty and Tushnet would like us to struggle for, is the state itself. Tushnet is correct in arguing that rights codified through the institutions of the state can be enormously destructive. They can suffer greatly from disutility; the wrong interests can be protected by rights. Moreover, the (capitalist) state is so fully complicit with the program of capital that it seems hopelessly utopian to think that it could ever be extricated and turned into a force of liberation.[23] That is precisely why so many on the left are willing to abandon state-centered approaches to social change (calling these approaches, rather than the state and capital, "totalizing") and substitute for them either the stern moralism that Rorty (1996) advocates, the cultural politics that he and others critique (Gitlin 1995; Tomasky 1995; see Kelley 1998), or the reliance on extraparliamentary, extrajudicial politics that Tushnet proposes in place of "rights-talk."

On the other hand, the (liberal) state[24] has proved itself—precisely through the institutionalization of rights—to be a key protector *of* the weak. These protections have not been freely given; they have been won, wrested through moralism, direct action, cultural politics, and class struggle, from the state and from those it "naturally" protects. Importantly, these fragile victories, incomplete as they are, counterproductive as they may sometimes be, are themselves protected only through their institutionalization in the state. To take only one example, Meghan Cope (1997) has argued persuasively that for women and children the state—the U.S. federal state at that—has been, in many ways, the best friend that they have had. The creation of a progressively democratic state (or even a first step toward that old dream of seeing the state wither away) must itself, in good part, begin by *strengthening the state*—*especially* in an era of "globalization." Put another way, the state is an essential player in contemporary capitalism and will remain so, no matter how much current political trends promote the *appearance* of its demise after Keynesianism (Meszeros 1995), the defeats of 1968, and the economic crises of the 1970s. To abandon the state to the forces of capital, or to those so efficient in organizing authoritarian populism (and political quiescence [Singer 1999]) *through* the state, is shortsighted in the extreme. "Rights talk" is one means by which the struggle to "capture" the state by progressives can be organized. "Rights" are one means by which progressive social policies can be instituted. Rights and rights talk, as the conservative legal scholar Robert Tier implicitly recognizes, are simply too important in the contemporary world to abandon in favor of some even more nebulous notion of morality (Rorty) or uninstitutionalized social struggle (Tushnet).[25]

What Rights Do

To make such a claim raises the obvious question; Just how do "rights" and "rights talk" do what I claim they do? For, while it is commonplace, it is also inaccurate, to assert that "discourse" *produces* things (like the social justice hopefully attendant upon socially progressive policies) Yet, this is not to say that discourses have no power. Quite the contrary, discourse helps set the context within which social practices occur and are given meaning. This power lies in the ability of words organized as discourse to *instruct*. Take the example of legal discourse. Laws and the discourse surrounding them can seemingly do all sorts of things. As laws, they can grant freedom or deal in oppression; they can order and regulate; or they can lead to mayhem. Yet, in reality, of course, it is not at all the legal words that do this. Words alone do not prevent striking workers in the United States from engaging in secondary boycotts;

words alone do not prevent (or allow) women to attend military school or engage in combat; words alone do not enable a corporation to take subsidies to locate its plants in certain communities—only to pull up its stakes a few years later, leaving in its wake a path of destruction. Rather, it is *police power*—the state-sanctioned threat of violence or other penalties—that permits these outcomes. At most, words can instruct and perhaps provide the discursive justification for the restraint or use of police power; they can help define other institutions of power that may or may not provide a check on the police power of the state. In this sense, words can provide a valuable tool for restraining power or for justifying it in particular ways. That is precisely what "rights" do: they provide a set of instructions about the use of power. But they do so by becoming *institutionalized*—that is, by becoming practices backed up by force (as Marx recognized).

Tushnet (1984, 1384) would counter that such an argument fails because of the indeterminacy of rights: "To say that rights are particularly useful is to say that they *do* something; yet to say that they are indeterminate is to say that one cannot know whether a claim of right will do anything." But Tushnet here ignores the way that a claim of right, *no matter how contested*, establishes a framework within which power operates. It matters less that power may breach this framework as often as honor it, because it is precisely in the breach that the political utility of rights talk does come to the fore. The abrogation of rights becomes a focus of political action, of social struggle. The argument (such as Tushnet would make) that claims of right cannot be determined within the discourse of rights is absolutely correct. The adjudication of rights, as Marx argued, is a function of force, a result of *political* action. It cannot be otherwise. But that does not thereby diminish the power of rights talk, as Tushnet claims it does.

For example, how would Tushnet, with his example of the need for food and shelter as pressing areas for social struggle, react to the widespread adoption of anti-homeless laws around the country (see Chapters 5 and 6)? Surely he would agree with Jeremy Waldron's (1991, 296) startlingly obvious assertion in this context that "no one is free to perform an action unless there is somewhere he is free to perform it." No matter how appalling it might be to argue and struggle in favor of the *right* to sleep on the streets or urinate in an alley, it is even more appalling, given the current ruthless rate at which homelessness is produced, to argue that homeless people should *not* have that right. That is, to the degree that we deny homeless people the right to sleep on sidewalks, we reinforce the "right" of the housed never to have to see the results of the society they are (at least partially) culpable in making. By denying the right to sleep, defecate, eat or relax *somewhere*, Waldron (1991) concludes, contemporary anti-homeless laws—predicated as they are on the *rights* of property—simply deny homeless people the right to *be* at all. In this instance, then, the denial that rights do anything (even if not autonomously) is genocidal. Likewise, absent the institutional power that rights talk helps to organize and constrain, it is hard to see how Rorty's call for compassion and moral persuasion will have any purchase against anti-homeless laws that take as their basis the twin "commonsense" notions that property rights must be protected and that there is no reason why people should urinate and sleep in parks and on streets.[26]

More directly to the point, moral arguments and compassion create no institutions that can protect moral and compassionate argument—at least not in a world still structured by states and their apparatuses. Rights, by contrast, have the force—physical, if partial—of law. What would happen if, in a few years and in the continued absence of enumerated rights to housing and livelihood, American society no longer produced homelessness quite so efficiently—that is, if homelessness became, because of reduced numbers, no longer quite so pressing a matter in most cities? How would the interests of those who remained homeless be protected when the attention of activists and advocates turned elsewhere, when moral and compassionate arguments were no longer quite so prevalent? Indeed, by what mechanisms can moral arguments protect a minority against a malicious (or simply selfish) majority?

It is helpful in this regard to understand the institutionalization of rights (or more generally the establishment of laws) as a moment in the production of space—especially material, physical space, not only the sort of metaphorical space that has been the currency of progressive theoretical development in recent years. While "production of space" theories are now quite complex,[27] it is in their barest outlines that one may best appreciate their utility for social movements seeking to create a just world. Lefebvre (1991) has argued that with the rise of capitalism came the hegemony of "abstract space." As labor was "abstracted" from social life (and as abstract labor came to dominate the social relations of production) under capitalism, *abstract* space was produced. This abstract space is different from "absolute space" (understood as a contentless container) because abstract space was and is socially produced under particular universalizing social relations.[28] As this abstract space becomes predominant, it becomes the site for the radical transformation of social life, establishing social life as merely a series of *abstract*, highly mediated, social relations (cf. Lukács 1968). Abstract space is the arrangement of space that makes capitalism possible, even as the social relations of capitalism make abstract space possible in the first place. Lefebvre (1991, 55) thus argues that "it is struggle alone which prevents abstract space from taking over the whole planet and papering over all differences."

The struggle for rights and for just laws is one aspect of this struggle to resist the hegemony of abstract space and to produce what Lefebvre calls "differentiated space." The struggle for rights is a determinate of the actual social content of the dialectic between abstract and differentiated space; the struggle for rights *produces* space. The rules for how capital moves across boundaries, for how firms develop in locations, for how public space is created, used, and transformed within cities are all, in part, rules of law, rules of *right*. Social action is structured through law, and social action creates abstract or differentiated spaces in proportion to the power possessed by each side in a struggle. So social action—including oppositional work by social movements—always operates simultaneously to influence the production of law and the production of space. As Blomley (1994a, 46) concludes, "law is, as it were, produced . . . in spaces; those spaces in turn are partly constituted by legal norms." The struggle for rights—for example, the right to sleep unmolested in a city park if you are homeless—becomes, as we will see in Chapters 3 through 6, an important, if still limited, tool in the production of space against powerful abstracting forces. But rights talk is more than a tool; if successful (and thus inscribed in law and policy), it provides institutional support for produced differentiated space to be maintained against the forces of abstraction that seek to destroy it. Rights themselves, therefore, are part of the process of producing space. Lefebvre (1991, 54) stakes out the end point of this argument: "A revolution that does not produce a new space has not realized its full potential; indeed it has failed in that it has not changed life itself, but has merely changed ideological superstructures, institutions or political apparatuses." The "cry and demand" for rights is a means for *producing* the right to the city—it *is* "that critique of human geography" Debord called for.

Space, Rights, and the Content of Justice

Any "critique of human geography" must be closely tied to normative philosophies of social justice. Without forging such a link, all the arguments by "little Arnolds" about order, freedom, and privilege will remain unchallenged in their claims to common sense. The critique of human geography, in the eyes of these "little Arnolds," is seen as a call for or an excuse for disorder, and thus dismissible. Linking social critique to social justice makes the invocation of "order" by those with material and property interests to protect less tenable: it can be shown to be the particular ideological move that it is. It shows that in fact the question is not one of order versus disorder but rather one of what sort of order is to be developed and advanced—a progressive one or a repressive and oppressive one.

The relationship between geography—the critique of human geography—and justice has been of keen theoretical, philosophical, and political interest at least since the early 1970s, with the publication of Bunge's (1971) report on the Fitzgerald neighborhood of Detroit and David Harvey's (1973) important analysis of liberal and Marxist urban theory, *Social Justice and the City*. In his book *Geography and Social Justice*, geographer David Smith (1994b) has since provided a very useful review of prominent philosophies of social justice that explicitly links them to contemporary geographical social theory. Reviewing both traditional (egalitarian, utilitarian, libertarian, contractarian) and more radical (Marxian, communitarian, feminist) theories of justice, Smith develops a sophisticated interpretation of egalitarian and distributive approaches to justice that takes space, place, and territory seriously. His goal is to illuminate the "structures responsible for inequality" (153) and to show that because such structures fundamentally concern questions of distribution (among other things) they are inherently geographic. To put that another way, Smith argues that exposing the geography of injustice is essential to developing social structures that are more just. Attention to geography forces a broadening of theories of injustice: it illustrates the ways that systems and structures of inequality become entrenched and reproduced in the actually existing world, and thus necessarily turns attention to questions broader than just distribution. Or rather, attention to just distribution within its geographical contexts demands struggle toward the transformation of those geographical contexts. Smith thus provides a sophisticated account of what is often called the distributive paradigm of social justice.

For Iris Marion Young (1990), reliance on a philosophy of distributive justice places such authors as Smith squarely in the mainstream of political philosophy.[29] "Contemporary philosophical theories of justice," according to Young (1990, 15), "tend to restrict the meaning of justice to the morally proper distribution of benefits and burdens among society's members." Moreover, such theories (not necessarily excepting Smith's) also tend to "stand independent of a given social context" while still claiming to "measure its justice" (Young 1990, 5). To do so, Young (1990, 7 *passim*) argues, contemporary theories of justice assume, within any given space, a homogeneous, undifferentiated, universal public, a public that shares like desires and needs. As soon as the universality and homogeneity of the public is revealed to be a myth, theories of distributive justice are exposed as inadequate to the task of rectifying *real*, socially situated *injustice*.

To make this argument, Young (1990) does not dismiss (as does Lyotard 1985) the contention that some universal normative claims can and should be made. For example, she suggests that "all reasonable persons share" the assumption that "basic equality in life situation for all persons is a moral value; that there are deep injustices in our society that can be rectified only by basic institutional changes; that [various social] groups . . . are oppressed; [and] that structures of domination wrongfully pervade our society" and can be dismantled (Young 1990, 14). Young's argument, in short, is that distributive justice (given the above universal assumptions) is vitally necessary[30]; but that it is not sufficient. Rather, the content of social justice must include, in addition to a just distribution of things, a framework that allows full, effective participation in decision-making by oppressed groups (35) and a frontal attack on various forms of oppression (48–63). "Justice," Young (1990, 37) argues, "is not identical with the good life as such. Rather social justice concerns the degree to which a society contains and supports the institutional conditions necessary for the realization" of two "values" essential to the construction of "the good life": "1) developing and exercising one's capacities and expressing one's experience, and 2) participating in determining one's action and the conditions of one's action" (Young 1990, 37, references deleted). As Young (1990, 37) recognizes, these are in fact *universal* values and they require their promotion for "everyone"; but they are also values that *demand* a careful attention to difference, for against them stand "two social conditions that define injustice: oppression, the institutional constraint on self-development, and domination, the institutional constraint on self-determination." Both oppression and domination are exercised through difference: it is difference that is oppressed and it is differently situated actors who dominate. Autonomy—the freedom to be who one is—requires not just the recognition of difference but also its social promotion.

In summary, for Young (1990) autonomy requires not simply a just distribution of goods and opportunities but social—or better, socialized—control over the *means* of distribution. And this socialized control has to be connected with elaborate, normative, universalizing, and institutional frameworks that promote autonomy and difference, both of individuals and of groups. Frameworks of *rights*, in other words, are crucial to the development of a social justice that moves beyond distribution and begins to recognize the struggle against oppression and in favor of autonomy (25). However, the ways in which we conceptualize "rights" needs to be transformed (96–97). Young argues, rightly, that within the discourse of law the " 'ethic of rights' corresponds poorly to the social relations typical of family and personal life" because such an ethic is based on a model of civic social relations that takes social detachment rather than social engagement as its basis.[31] Critiquing the Habermasian ideal of a detached "public sphere" and drawing on a range of feminist arguments, Young (1990, 97) notes that the "ideal of impartial moral reason" (which stands behind much rights talk) "corresponds to the Enlightenment ideal of the public realm of politics as attaining the universality of a general will that leaves difference, particularity, and the body behind in the private realms of family and civil society."

Such a conception of rights and with it, such a conception of dispassionate social justice—relies on what Young (1990, 98) calls a "logic of identity" that "denies or represses difference." This is because "the logic of identity tends to conceptualize entities in terms of substance rather than process or relation." But a more dialectical notion of entities (see Ollman 1990; Harvey 1996) can be adopted, struggled for, and defended. Doing so would mean that the "logic of identity" has to be replaced with a "logic of representation." A "logic of representation" centers on the right of groups and individuals to make their desires and needs known, to represent themselves to others and to the state—even if through struggle—as legitimate claimants to public considerations. Such a logic requires the acceptance of a (near) universal and positive *right* of representation. Yet, as with any other right, such a right cannot be guaranteed ("accepted") in the abstract—rather, it is something always to struggle toward. In this struggle, the development—or often the radical claiming—of a *space for representation*, a place in which groups and individuals can make themselves *visible*, is crucial. While it is no doubt true that the work of citizenship requires a multitude of spaces, from the most private to the most public, at the same time public spaces are decisive, for it is here that the desires and needs of individuals and groups can be *seen*, and therefore recognized, resisted, or (not at all paradoxically for thoroughly materialist rather than idealist normative social practices) wiped out. The logic of representation demands the construction—or, better, the social production—of certain (though not necessarily predetermined) kinds of public space.

Representation and Public Space

Representation, whether of oneself or of a group, demands space. While it is true that "human beings have no choice but to occupy a space: they just do," as David Smith (1994b, 151) puts it, it does not follow that such a space allows for the full, adequate, and self-directed representation of human beings either to themselves or to others. Jail and school, to take two obvious examples, are controlled environments where the needs, desires, rights, and hence ability to self-represent are not only limited but often denied altogether. But one does not have to focus on such "total institutions" (as Foucault has called them) to see that the ability for (or right of) representation is continually frustrated everyday for innumerable individuals and groups. While occupying some place or space is vitally necessary to life, it is not necessarily guaranteed as a right.[32] Rather, private property rights hedge in space, bound it off, and restrict its usage. As Smith (1994b, 42) argues, "the right to own land differs from other commonly enunciated rights, in that it concerns the appropriation of the scarce material world, and can impinge on the rights of others to meet such vital needs as food and shelter." Moreover, private property rights also potentially trump what Smith (1994b, 43) calls membership rights but which in the American context might be more commonly understood as the right to assembly—that is, those rights that make possible the formation of political communities, that make possible political *representation*.

In a world defined by private property, then, *public space* (as the space for representation) takes on exceptional importance. At the level of basic needs, as Waldron (1991) argues, in a society where *all* property is private, those who own none (or whose interests aren't otherwise protected by a right to access to private property) simply cannot *be*, because they would have no place to *be*.[33] At a less immediate but still vital level, in a world defined by private property, the formation of a *public sphere* that is at all robust and inclusive of a variety of different publics is exceedingly difficult.

Much ink has been spilled arguing the merits of Habermas's (1989) notion of the public sphere, and especially the limited historical geography out of which he saw it arising (cf. Calhoun 1989). As numerous critics have pointed out, Habermas's singular and normative theory of the universal public sphere is handicapped from the beginning because it attempts to universalize a model of discourse that developed in highly constrained, exclusive (male, bourgeois, white) spaces, such as the 18th-century coffee house. Nancy Fraser (1990) argues that the notion of the singular universal public sphere needs to be replaced with a theory of multiple, contending, often mutually exclusive public spheres. Just as important is the need to provide a more realistic geographical basis to the very notion of the "public sphere."

Implicit in much theorizing about the public sphere is the assumption that the provision of an adequate space (or in some renderings, an adequate technology) will perforce create a vibrant public sphere.[34] The proliferation of and perhaps democratic control over places to meet, gather, and interact (whether these places be town squares, electronic communities, televisions chat shows, or "the media") are often seen as sufficient to the creation of a public sphere. The erosion of such places (and their replacement by privately controlled spaces and means of communication) is likewise often argued to be crucial to the closing down of the public sphere. The images—or ideals—of the public sphere and its relationship to space are important and in their normative force often drive much political organizing and action. And yet these arguments are limited to the degree that they assume that the construction of either singular or multiple public spheres is an issue of planning, and that such planning is—or could be—sufficient to the promotion of political discourse. The planning and provision of public spaces will lead, the argument often goes, to the ability of various groups to represent themselves.

And yet, as careful analyses of the community network movement in the United States (such as that by Michael Longan 2000) show, even the most well designed spaces for interaction (in this case the electronic space of the internet) often lead to a remarkably limited and ineffectual public discourse. Indeed, Longan (2000) found that the most effective arenas of public discourse arose around specific issues and specific needs. That is to say, political debate developed not because it *could*, but because *it had to*—and in the process often the least likely sites for political representations became the most important.

That is to say, for all the importance and power of recent "end of public space" arguments (which is great, as I will discuss in Chapter 4), what makes a space *public*—a space in which the cry and demand for the right to the city can be seen and heard—is often not its preordained "publicness." Rather, it is when, to fulfill a pressing need, some group or another *takes* space and through its actions *makes* it public. The very act of representing one's group (or to some extent one's self) to a larger public creates a space for representation. Representation both demands space and creates space.

But it rarely does so under conditions of its own choosing. And so here the desires of other groups, other individuals, other classes, together with the violent power of the state, laws about property, and the current jurisprudence on rights all have a role to play in stymieing, channeling, or promoting the "taking" and "making" of public space and the claim to representation. The move again and again to Hyde Park—or to create new Hyde Parks—meets deep opposition, not only from innumerable "little Arnolds" but also from riot police wielding tear gas, corporate lawyers wielding writs and subpoenas, and "rent-a-cops" wielding revolvers (and licensed to use them). So too is "Hyde Park" reclaimed by the almost inevitable attrition endemic to any militant or long-term occupation—whether that occupation is one of homeless people creating a communal shantytown on city property or the occupation of generations of activists seeking to retain the right to protest in (or just to use) spaces subject to the imperial designs of corporate capital and its allies in planning and land-use departments.

The production of public space—the means through which the cry and demand of the right to the city is made possible—is thus always a dialectic between the "end of public space" and its beginning. This dialectic is both fundamental to and a product of the struggle for rights in and to the city. It is both fundamental to and a product of social justice (which thus cannot be universal *except* to the degree it relates to the particular and the spatial—particular struggles for rights and particular struggles over and for public space). The purpose of the chapters that follow is to explore—in historical-geographical detail as well as at the level of normative theory—just this dialectic, and to show how it structures the "right to the city" as it actually exists and as various activists and social groups have struggled to make it be.

NOTES

1. The full sentence reads: "The rough has not quite found his groove and settled down into his work, and so he is just asserting his personal liberty a little, going where he likes, assembling where he likes, bawling as he likes, hustling as he likes" (Arnold 1993, 88). The invocation of settling down into one's appointed work is telling. But more important, perhaps, is that Arnold makes his argument as a defense of the State (the capitalization is his), which he sees as both the guarantor of order and the (perhaps imperfect) expression of perfection. This sets Arnold apart from many of the contemporary "little Arnolds" writing in America whom we shall shortly meet.

2. My argument throughout the course of this book will *not* be that "order" in and of itself is bad; rather, the issue is what kind of order, and protecting whose interests, is to be developed and advanced. Indeed, I will argue, especially in the Conclusion, that "order" is as vitally necessary to the progressive city as it is to the oppressive or repressive one.

3. In late 1999, responding to a highly publicized assault that was wrongly linked to a homeless street person, Mayor Rudy Giuliani of New York reiterated his (and many others') staunch belief that there simply is "no right to live on the streets." Giuliani put it starkly: "Streets do not exist in civilized societies for the purpose of sleeping there. Bedrooms are for sleeping in" (Bumiller 1999, A1)—which, of course, is fine if you have one. For those who do not, Giuliani announced a new program to arrest those sleeping on the streets if they did not "move on" when ordered to do so by the police. Simultaneously, Giuliani announced that shelter beds would be conditional on employment. Most of the homeless, under this policy, were caught in a quite sharp "Catch-22." As the *New York Times* put it in an analysis, "many New Yorkers seemed puzzled by a policy that would throw

homeless people out of shelters and into the streets, and yet arrest them for being there if they would not go to a shelter" (Bernstein 1999, 1). Indeed.

4. Here Will is quoting the editor of the conservative *American Enterprise* magazine, Karl Zinsmeister, who argues that those who promote homeless people's rights to public space "have no regular experience of neighborhood life as it must be endured by low-income city dwellers" (quoted in Will 1995). Neither Zinsmeister nor Will provide any evidence for this allegation.

5. The Manhattan Institute has produced a compendium of polemic from the urban right (Magnet 2000), which fairly clearly outlines the current "little Arnold" position.

6. The AARR is a branch of something called the "Communitarian Network," both of which are dedicated to litigation that they hope will "restore the spirit of community in the United States" by striking "a balance between extreme rights claims and those who would sacrifice civil liberties as a means to an end" (Golden 1998, 552). Tier is now the President of the Center for Livable Cities, "a national non-profit organization that assists local governments, downtown associations, and citizen anti-crime groups on urban crime and disorder issues" (Tier 1998, 255n).

7. See note 3, above.

8. The "Million Youth March," held in Harlem on September 5, 1998, was organized by Khalil Muhammad, a leader of the New Black Panther Party and a former leader of the Nation of Islam (who had been expelled by Louis Farrakhan for being too bigoted). The rally was required, by order of Mayor Giuliani, to end at 4 P.M. When it went overtime by a few minutes, riot police stormed the stage, sparking a riot. A month and a half later, some 4,000 New Yorkers gathered to march through Manhattan in memory of the brutally murdered gay University of Wyoming student Matthew Shepard. Police reacted by chasing marchers through the streets with nightsticks.

9. Giuliani's anti-community gardens campaign has received a huge amount of attention, and yet there is not yet a single good source for understanding its history. What is clear, however, is the degree to which, prior to September 11, 2001, his alienation of much of his political base through this campaign had undermined his support and popularity. The problem that faced Giuliani when he turned against the gardeners (largely, it seems, out of an ideological commitment to private property and a visceral dislike of populist movements that had been successful at implementing what he desired to implement through more authoritarian means) was that he was leading an assault against one of the things that New Yorkers most liked rather than disliked about their city.

10. Reviews of Giuliani's quality-of-life policies and practices may be found in N. Smith (1996, 1998).

11. For a discussion of the relationship between these two books, and for an

argument that Lefebvre's deeply abstract arguments were in part a function of his style of work, which relied heavily on dictation, see Shields (1998). The best examination of the development of a specifically urban Marxism in Lefebvre's work is now to be found in the chapter on Lefebvre in Andy Merrifield's (2002) wonderful new book *MetroMarxism*.

12. There are, of course, more basic arguments as to why the city must be at the heart (but not at all the exclusive focus) of any struggle for a progressive, socially just world. Among these are the simple fact that most of the world's population is now urban, that cities have become the command and control centers of the global economy and of the practices and policies that are transforming the global environment, and that, in fact, increased rather than decreased urbanization will have to be at the heart of any move toward sustainability under continued population growth: cities are every bit as much a solution as they are a problem.

13. The Detroit Geographical Expedition and its offshoots in Toronto and elsewhere marked an important moment in the rise and fall of an *activist* community-centered geography that saw its task as one of helping to develop the theories and skills necessary for radical, bottom-up social transformation. The impressive growth of social theory in the past two decades has come at the cost of abandoning this project, no matter how much radical geographers have retained the language of activism. Merrifield's (1995) examination of Bunge's geographical expeditions does a superb job of both recovering the history of the expedition (including its flaws) and charting this shift in activist geography.

14. Lyotard's (1985) broadside argues that "even the emancipation of humanity" (60) is disqualified as a grounds for universal discourse, and yet it concludes with an explicit discussion of justice "as a value [that] is neither outmoded nor suspect" (66).

15. The best discussions of the geography of rights are authored by Nicholas Blomley (1994a, 1994b).

16. All this is to say (to put it bluntly), "rights" must be at the heart of any Marxist and socialist project of urban transformation, even while the limits of rights, and the need to continually struggle over them, must constantly be acknowledged.

17. This and the next section are revised versions of arguments I first made in Mitchell (1997a).

18. Negative rights are expressed as a prohibition on some aspect of state power; for example, a prohibition on the state's interfering in people's speech acts is negative rather than positive because it defines the limits of state action rather than the extent of people's ability to speak.

19. His primary example, reflecting a set of court cases of the time, was the assault on the rights of gays and lesbians, and in this context he argues that such an assault will not be stopped by justices discovering "a hitherto invisible right to sodomy" but because the heterosexual majority will be persuaded that "tormenting homosexuals for no better reason than to give itself the sadistic pleasure of humiliating a group designated as inferior" is

bad. Of course, such a remedy says nothing about how to redress grievances by those gays who continue to be "tormented" (or even just discriminated against) even after the majority has come around.

20. Those of us not well-versed in philosophy might be led to wonder why "moral discourse" is more practical and concrete than the language (and institutionalization) of rights, but, unfortunately, Rorty does not provide an answer. Nor does he provide an answer as to why progressives cannot fight for economic as well as cultural rights.

21. Authoritarian populism probably better describes the ruling ideology of Thatcher/Reagan than Clinton/Blair. The zeitgeist of the 1990s might be better characterized as "therapeutic authoritarianism." Whatever populist energies Thatcher and Reagan may have galvanized were dissipated by Clinton and Blair, who both so readily "felt our pain" and then increased it by ending "welfare as we know it" and instituting rather draconian disciplinary compulsive-employment laws in its place. The authoritarianism remains, but now it is made palatable by adding to it a warmed-over mixture of 1970s-style psychobabble and '80s-style "tough love." As I write, in the wake of September 11, 2001, in the midst of the U.S. and British war against Afghanistan, and as the anthrax scare continues to remain unsolved, both President George Bush and Prime Minister Tony Blair seem intent—with their stepped-up police powers for unwarranted searches, proposals to read private e-mails and listen in on attorney–client conversations, and mass detentions of the uncharged and unarraigned—to create a new fascism, a sort of (nominally) democratic, authoritarian totalitarianism. It is no accident that this is being put into place through a wholesale attack on rights—the right to privacy, to be free from unreasonable search and seizure, to a speedy trial and to confront your accusers, and so forth.

22. Blomley (1994b) makes a compelling argument for the importance of "rights-talk"—an argument that has been a deep inspiration for my own thinking on the subject. Following Patricia Williams (1991), Blomley rightly notes that the leftist critique of rights, which has been conducted largely by whites, is quite condescending when the everyday experience and ongoing struggles of African Americans is considered. For African Americans, as for other people of color, many women, the homeless, and sexual minorities, and now Arab and Islamic peoples, everyday experience features the routine denial of formally recognized rights. The struggle for the effective protection of such rights, therefore, is central to progressive organizing in these communities. Harvey (2000) develops this argument in specific reference to the important promissory power of the United Nations Declaration of Human Rights.

23. As Daniel Singer (1999) has made clear, ideologues for capitalism have used this sense of the impossibility of change to great advantage in the past several decades. Summarized by Prime Minister Margaret Thatcher's famous phrase "There Is No Alternative," this position argues that global capitalism and continued oppression are not just the best possible alternative, but rather the only possible alternative.

24. The liberal and the capitalist state are, of course, one and the same. The state is contradictory (Clark and Dear 1984; Habermas 1974).

25. I say "one means" because there are clearly others: both progressive and reactionary economic policies can be enshrined in law, for example, as can many moral principles (also both progressive and reactionary), but at root these must entail an assessment of rights: both what is right and that to which people are entitled only by dint of their residence in a political community. All of these terms—rights, laws, political community, residence—are indeterminate and therefore open to ongoing struggle. This is why discussions of *power* are vital to any discussion of rights, morality, and so forth. The difference between morality and uninstitutionalized social struggle on the one hand, and rights on the other, is that the latter *necessarily* demands a theory of power—and its exercise—at its heart. Discourses of morality assume human perfectability; uninstitionalized social struggle assumes that there never are winners who can organize power and violence to their overwhelming advantage; the struggle for rights and their institutionalization at least frankly admits that some people are shits and will do all they can to screw others unless restrained from doing so by the full institutional and police weight of society.

26. The 1990s discovery of "compassion fatigue" among the urban and suburban middle and upper classes has at least in part licensed the remarkably mean-spirited legal attack on homeless people during that decade. Beyond its inherent paternalism, as a force for progressive change "compassion" has very clear limits.

27. Good discussions of the literature may be found in Blum and Nast (1996); Brenner (1997); Gregory (1994); McCann (1999); Merrifield (1993, 2002); N. Smith (1990); Soja (1989).

28. This argument is fully developed in Ch. 3 of N. Smith (1990).

29. D. Smith (1994b, 103–107) spends several pages developing and critiquing Young's (1990) arguments about the relationship between justice and the politics of difference. Smith concludes (1994b; 107) that Young's theory of justice is ultimately "Utopian," but as we will see, when it is placed in *its* geographical context such a charge cannot be sustained. Young's work has received a great deal of attention in geography. See Harvey (1996, Ch. 12) and Minnesota Geography Reading Group (1992).

30. "The immediate provision of basic material goods for people now suffering severe deprivation must be a first priority for any program that seeks to make the world more just" (Young, 1990).

31. Michael Brown (1997) and Lynn Staeheli (1994) have begun to explore the complex geography of the "work of citizenship." In Brown's case, primary inspiration is taken from Laclau and Mouffe's (1985; Mouffe 1992) poststructuralist development of "radical democracy" that suggests that the moment of democracy may or may not be easily "public" in any traditional sense. Staeheli develops feminist arguments (e.g., Pateman 1989) to make the same argument as she shows the widely varying locations of women's political work.

32. See Chapter 6 for a discussion of how the California Supreme Court relied on this point to deny homeless people a claim to the right to sleep in public when no other housing was available.
33. We will examine the degree to which such a world is being constructed in the United States in Chapters 5 and 6.
34. "Public sphere" is a complicated term in that it indicates public interaction at a range of scales and across several levels of abstraction. Both people gathering at a town meeting and the sum total of discourse in the media are often referred to as aspects of the public sphere. My argument is not that these are not constituents of public spheres but rather that much academic debate about the public sphere occurs amidst a high level of analytical imprecision.

2

Making Dissent Safe for Democracy

Violence, Order, and the Legal Geography of Public Space

The right to the city is never guaranteed and certainly never freely given. Indeed, it is never, a priori, clear to whom the right to the city belongs: that too is decided in the crucible of social struggle, struggle that ranges from the home to the city streets and, in the United States, into the chambers of the Supreme Court. That is to say, the "actually existing" right to the city—and the struggle for its expansion by some social groups and its contraction by others—is the product of specific social contests, in specific places, at specific times. And yet, these contests themselves give rise to particular forms of regulation—or the adjudication of interests—that, as *law* or as formally enunciated rights, are universal (or better yet, universalizing) in their intent.

Take two recent decisions by the United States Supreme Court, two decisions that are particularly complex for those of us interested in promoting a *progressive* right to the city. Both concern the rights of women seeking abortions as they are confronted by protesters outside abortion clinics. The first established the need for especially sensitive laws regulating the *conduct* of protesters (*Madsen v. Women's Health Center* 1994). The second concerns one such law and its constitutionality (*Hill et al. v. Colorado et al.* 2000). Taken together, the two suggest a certain way of thinking about public space, and hence the right to the city. But both also indicate a particular understanding of public space that is deeply problematic, even as it has been developed out of a long history of social struggle and judicial regulation, a history that both deeply influences *how* people struggle to claim a right to the city *and* how they understand the relationship between social practice and political ideas (and ideals). The purpose of this chapter is to work backwards from these two cases to understand both the logic governing judicial (and therefore police) regulation of public space and the history of struggle that gave rise to that logic. For it is in this history that we can begin to glimpse not only how people struggle in public space for their rights but also how the content of public space itself conditions that struggle. We can also see how protest in one realm of social life—the struggle between capital and labor, for example—can have profound effects on protest in another—the struggle between those who fight for a woman's right to abortion and those who would deny it. Such effects, of course, are what give the "cry and the demand" for the right to the city its poignancy and its force.

BUBBLE LAWS, ABORTION RIGHTS, AND THE LEGAL CONTENT OF PUBLIC SPACE

Decided in 1994, *Madsen v. Women's Health Center* announced the constitutionality of an important regulatory strategy for controlling violence by anti-abortion protesters. Ruling to strike down some parts and uphold others of a lower court's injunction regulating protests outside an abortion clinic in Melbourne, Florida, the United States Supreme Court declared that the streets around the clinic constituted a "traditional public forum." Any restriction of "speech" in this forum had to be accorded the highest scrutiny to assure that it was content neutral (*Madsen* 1994, 2524)—that it did not revolve around regulating *what* people were saying, only how (and when and where) they might say it. The Court ruled that some regulation of anti-abortion protesters was required because protesters' *actions* (rather than what they were actually saying) both threatened violence and threatened the rights of others. The Court therefore upheld provisions in the injunction that barred protesters from occupying a 36-foot buffer zone around the clinic entrances and driveways (excluding private property that could not be regulated as a public forum). The level of noise made by the protesters could also be regulated. However, the court ruled that a 300-foot zone around the clinic in which the protesters could not approach employees or patrons was too great a burden on the free speech rights of protesters: a much smaller exclusion zone would have to be drawn. Similarly, a ban on signs and banners together with a 300-foot restricted zone in the streets near clinic employees' homes were also invalidated (*Madsen* 1994, 2526–2530). In short, drawing on a large body of precedent, the Court held that it was within the power of lower courts to establish appropriate "time, place, and manner" restrictions on the exercise of speech and assembly by anti-abortion protesters (*Madsen* 1994, 2524).

Partially in response to this decision (which was geared toward a specific injunction but which established the broad conditions under which anti-abortion protest outside clinics and clinic employees' homes could be regulated), a number of state legislatures and local jurisdictions created "bubble laws" regulating the actions of protesters within certain distances of clinics and homes. In addition to regulation of the fixed space around entrances to clinics, many such laws also established a protective, moving "bubble" (typically 8 to 15 feet) around patrons and employees entering clinics. It was one such law that the Supreme Court upheld by a 6–3 majority during the 2000 Court term (*Hill* 2000). Passed by the Colorado state legislature in 1993, this law specifically held (in part) that it is a misdemeanor if a person "knowingly obstructs, detains, hinders, impedes, or blocks another person's entry or exit from a health care facility" (Colo. Rev. Stat. § 18-9-122 para. 2). The "bubble" restriction was quite specific:

> No person shall knowingly approach another person within eight feet of such person, unless such other person consents, for the purpose of passing a leaflet to, displaying a sign to, or engaging in oral protest, education, or counseling with such other person in the public way or sidewalk area within a radius of one hundred feet from any entrance door to a health care facility. (C.R.S. § 18-9-122 para. 3)

As with the more general restriction, anyone violating this "bubble" provision is guilty of a misdemeanor.

In his opinion for the majority, Justice John Paul Stevens noted that while the law made it more difficult to hand a leaflet or otherwise "give unwanted advice" to someone entering a clinic, it did not entirely prohibit anti-abortion protests. While the law prohibits protesters from "approaching" people, "it does not require a standing speaker to move away from anyone passing by" (*Hill* 2000, 2485). In its clearest statement, the majority opinion asserted that the Colorado statute "places no restrictions on—and clearly does not prohibit—either a particular viewpoint or any subject matter that may be discussed by a speaker. Rather, it simply establishes a minor place restriction on an extremely broad category of communication with unwilling listeners" (*Hill* 2000, 2493).[1] That is, the statute regulates *space* and *spatial relationships* as a means of protecting a woman's right of access to abortion (and other health) clinics free from interference by those who would wish her not to seek an abortion. The majority held that this law did not restrict the *content* of anti-abortion protests because it no less prohibited *all other* messages within the bubble established by the law: "the statute applies equally to used car salesmen, animal rights activists, fundraisers, environmentalists, and missionaries" (*Hill* 2000, 2493).

While the majority is certainly correct in its opinion, the fact remains (as the majority freely admits) that the legislative history of Colorado's law makes it clear that anti-abortion protesters were the specific target of the law—not used car salesmen, fund-raisers, or animal rights activists. In Colorado, as in many other states, anti-abortion "sidewalk counselors" had been relatively effective in both deterring women seeking abortions from entering clinics and intimidating clinic workers. Indeed, links were made, both in the public mind and in courts of law, between sidewalk counseling and even more violent forms of anti-abortion protest, such as the bombing and torching of clinics, harassment of and assassination attempts on abortion clinic workers, stalking, and so forth. Colorado's law, like the later Federal Access to Clinic Entrances law, was designed to redress some of these problems by regulating the space around a clinic entrance so as to better assure that a woman's right to abortion was more than just a "paper right"—that it was a right that could be practiced.[2] The *content* of the protesters' messages, as well as the protesters' conduct, was exactly what was at stake—no matter how much the majority argued to the contrary.[3] Indeed, this is exactly why this case, and others like it, is important. Under U.S. constitutional law, "political" speech is afforded a great deal of legal, and even more ideological, protection. The Supreme Court professes itself to be keenly interested in assuring that the state does not "overly burden" the right to political speech and often goes to impressive lengths to find ways to regulate or even ban certain kinds of speech (and certain messages too) indirectly while at the same time claiming to *protect* the very speech it is prohibiting. Perhaps the primary means of accomplishing this goal is through the regulation of space itself. The "right to the city" therefore is always vetted through, and to some degree regulated by (even if only in the negative), a geography of law. In turn, the very nature of urbanism is at least in part a product of the struggle over the legal content of public space—who owns it, who controls it, who has the right to be in it, and what they may or may not do once there.

Seeing public space in these terms allows us to understand that "law" is always "enacted," even performed, in particular spaces and at particular times. Law, and the "rights talk" that accompanies it, shapes social struggle in particular ways—as for example when activists shape their protest strategies in particular public spaces so as to conform to or actively resist specific laws and regulations. But so too do these enactments and performances stretch and sometimes even break the law, requiring the pieces to be reassembled, perhaps by the Supreme Court, perhaps by lower courts, or perhaps, as we will see below (Chapters 5 and 6), by crusading city administrations and police forces, into a new whole, a new set of rules and regulations that seeks to emplace universalizing order on specific public spaces. A wide range of scholarship in geography (e.g., Blomley 1989, 1994a, 1994b, 1998, 2004; Blomley and Clark 1990; Bakan and Blomley 1992; Chouinard 1994; Clark 1990; Delaney 1998, 2001; Pue 1990; in general see Blomley, Delaney, and Ford 2001) has explored the contradictions that arise because laws (and rights) are simultaneously situated in specific spaces and places—specific spaces such as the sidewalk in front of a women's health clinic, give rise to specific regulatory and legal strategies—and universalizing in that they either cover much wider spaces or are easily transferable to different sorts of spaces. As both the majority and the dissent in the Colorado bubble case recognized, the law in question could easily be adapted to a range of other circumstances and a range of different kinds of space—just so long as there was a "substantial state interest" in regulating particular behaviors and kinds of speech. The purpose of this chapter is to further explore this contradiction through the lens of a particular complex of laws originating in a set of cases seeking to regulate speech that was feared to be "seditious," transferring into a set of cases that sought to tightly regulate labor protest, and coming to full flower in the abortion clinics cases of *Madsen* and *Hill*. This complex of laws seeks to regulate both the content and the form of protest by regulating not protest itself but the space in which that protest occurs (and thus which is produced, at least in part, *by* protest).

REGULATING PUBLIC SPACE

In the Colorado bubble law case, the Court drew heavily on the earlier *Madsen* decision that upheld the constitutionality of an injunction against anti-abortion protesters in Florida and thus paved the way for the dozens of local, state, and federal laws like the Colorado one that enshrined both stationary and moving exclusion zones around women seeking access to clinics in the statutes. In *Madsen* the Supreme Court attempted to delineate a space of protest that retained the protesters' rights while "ensuring safety and order . . . promoting the free flow of traffic on public streets and sidewalks, and . . . protecting the property rights of all citizens" (*Madsen* 1994, 2526, citing *Operation Rescue v. Women's Health Center, Inc.* 1993, 672). The clear assumption of the Court, and one that, as we will see, is well-developed in nearly a century of Supreme Court decisions, was that a public space of protest could only "work" if it was orderly and free of potential violence (to people or to property). The Court reasoned that tight but reasonable boundaries had to be drawn around the public space to retain it as a place open for public political activity. Only in this way could *peaceful* protest be promoted while the rights of clinic patients and employees were also protected. In some ways, the Court's decision was a classically liberal one that assumed that ideas and speech acts could be readily divorced from actions—from their very performance—and still retain their power. Such a line of reasoning argues that a democratic polity requires dissenting ideas; these ideas, however, have to stand or fall on their own merits as they enter into competition with other ideas; the better ideas win, but only by being tested against less worthy ideas. Speech, the act through which ideas are put into circulation, must thus be protected in a democracy. If, however, speech is *forced* into the public arena, or if violence accompanies that speech, it can and must be regulated. The assumption behind this ideology, of course, is that all individuals have equal access to what the liberal Supreme Court Justice William Brennan once famously called "the marketplace of ideas" (*Lamont v. Postmaster General* 1965).

For the Supreme Court, at least since the 1930s, the place where ideas come into competition with one another has been "the streets and parks" and other public forums that "have immemorially been used for the purpose of assembly, communicating thought between citizens, and discussing public questions" (*Hague v. CIO* 1939, 515). Public space is imagined to be the site of political inclusiveness, a place in which interested individuals can come together to debate and to voice dissent. According to the democratic ideology enshrined in what the Court has come to call "public forum doctrine," the free exchange of ideas can occur only when public space is orderly, controlled (by the state or other powerful interests that can maintain order), and safe. The free exchange of ideas is predicated on Arnoldian order. The goal for the Court, or at least those members seeking to uphold classically liberal notions of democracy, therefore has been one of writing rules for public space, rules that will make dissent possible, but possible *only* if it can be shown to be free of "force" or "violence."

The public forum doctrine upon which *Madsen* and *Hill* are based seeks to define what Kalven (1965) called a "new Roberts Rules of Order" for public space.[4] These rules of order are designed to protect the "marketplace of ideas" from *conduct* that is threatening to the exchange of ideas. The Court has created three types of public forums, each of which must be regulated differently. The first type—those streets, parks, and other places that have been open to the public "from time immemorial" (*traditional* public spaces)—must have the least restrictive laws governing the exercise of speech. Any laws regulating the "time, place, and manner" of speech and assembly must be rigorously "content neutral." The second level of space is *dedicated* public space—such as plazas in front of federal buildings or portions of public college campuses—which a government or state agent has designated as open to public speech and assembly. Like traditional public forums, these spaces can be regulated in terms of time, place, and manner, but not in terms of speech content. But unlike traditional public forums, the state has specifically enabled these spaces and thus has a greater right to remove them from public use. The final form of public forums includes all remaining public property.[5] In this class (which includes everything from open-space trails and undedicated plazas to schools and military bases) speech activities can be freely regulated (Haggerty 1992/1993, 1128).

In the *Madsen* case, the Court considered the streets around the abortion clinic in Melbourne, Florida, to be a traditional public forum, but one which, because of the *conduct* (rather than the speech content) of the protesters, had to be carefully regulated. Justice Antonin Scalia did not think that the Court did a very good job of regulation. In his vigorous dissent, Scalia argued that the majority of the Court left "no doubt" that their goal was to "restrain persons not by proscribable *conduct*, but by proscribable *views*" (*Madsen* 1994, 2540, emphasis in original). By restricting the conduct of only anti-abortion protesters and not prochoice counterprotesters (whom Scalia described as just as disorderly and potentially violent, if not more so), the Court clearly had eliminated the right of dissent in public space. "[T]he Court," he concluded, "has left a powerful weapon lying about today," waiting to be used by those who are not "friends of liberty" in their attempts to restrict the public speech and assembly rights of militant groups (*Madsen* 1994, 2549–2550). Using an "ad hoc nullification machine" that seemed to work only in cases concerning abortions, Scalia complained, the Court had claimed "its latest, greatest, and most surprising victim: the First Amendment" (*Madsen* 1994, 2535). Scalia repeats exactly this argument in his dissent from *Hill* (2000, 2503), arguing further that under the Court's doctrine the First Amendment is now pretty much "a dead letter" (2507). Indeed, according to Scalia, "[t]here is apparently no end to the distortion of our First Amendment law that the Court is willing to endure in order to sustain this restriction upon the free speech of abortion opponents" (2508).

The *Madsen* and *Hill* decisions are an important if uncomfortable place to begin an historical geography of the legal structure of public space because, contra Scalia, the decision of the majority in both cases is in general accord with developing doctrine concerning the use of public space. And this doctrine, as we will see, is both formative of and troubling to any ideological notions we may unreflexively hold about public space as a center of democratic discourse. The right to the city is in part, vetted through decisions like those made by the Court in *Madsen* and *Hill*. The restrictions that Scalia sees as a "powerful loaded weapon" ready to slay cherished First Amendment ideals are really nothing new and in fact merely confirm a general trend of spatial regulation inaugurated by the Court in the early 20th century as a means for making dissent safe for democracy. Standard legal histories of free speech and assembly write that history as one of progressive enlightenment in which onerous old restrictions on public assembly and speech have been swept away as successive Courts have realized the value of dissent to the democratic project (for a review see Sunstein, 1992). According to this history of progress, the real liberalization of public space law begins with the ringing dissents of Justices Oliver Wendell Holmes and Louis Brandeis in the 1910s and 1920s as they fought a conservative Court majority bent on restricting nearly all dissent in the name of order. By the 1930s a more liberal Court majority finally ratified the opinions of Holmes and Brandeis, and the promise of the First Amendment was at last fulfilled. The problems of First Amendment law since then have been technical: how does the state write regulations that simultaneously protect speech and maintain order?

The point of the remainder of this chapter is to show how this history of progressive enlightenment, if confined to the level of Court decision making (as it often is), is both too simple and incapable of explaining either the nuances of decisions such as *Madsen* and *Hill* or the tortured politics of Scalia's dissents. Nor is it capable of shedding much light on the absolutely critical importance for understanding struggle over the right to the city that nonetheless attaches to the work of such regulatory bodies as the Supreme Court and their ability to make and define law. The history of progressive enlightenment also makes it difficult to see just why it might be necessary, as Raymond Williams says, for people to "go again to Hyde Park"—to constantly meet and speak in public space not only for some particular issue but also to claim again and again the very right to meet and speak in public space.

The way to understand these issues is to uncover the source of political and regulatory change in the actual battles that have led to the formulation of public forum doctrine (or other regulatory apparatuses). By doing so we can better grasp how judicial and legislative lawmaking works dialectically with social and political action to structure public space itself. One of the shortcomings of traditional legal analysis, as Nicholas Blomley (1994a) has shown so clearly, is that it takes the universality of law for granted when it should show how that universality is built out of particular spaces and particular times and is thus responsive to those spaces and times in ways that may make it inappropriate for other social contexts. I will suggest in the following sections that, by looking at the recursive interactions of law and social action in particular urban "public" spaces, we can begin to see how so basic a judicial tool as public forum doctrine developed not so much as a means for assuring First Amendment rights (as Justice Scalia assumes in his dissents) but rather as a tool for making certain kinds of dissent possible for a certain *kind* of democracy. The Court, as we shall see, has been both a *site* of progressive change (though a quite limited one) *and* a force for protecting, in the name of order, dominant interests against the very kinds of acts that make dissenting speech possible. Indeed, the history of public forum doctrine, as we will see, begins with a desire by lawmakers and the judiciary to control dissenters, particularly radical workers seeking to redress political and economic injustices.[6] The major difference between the abortion cases and earlier cases involving labor picketing (cases that are cited with great frequency by both sides in *Madsen* and *Hill*) is mostly that the state's interest in regulating dissent has moved from radical labor to radicals of other sorts. The question *Madsen* and *Hill* thus raises is not just one of public forum doctrine as it relates to real spaces and real actors, as it structures public space, but also one of the *adequacy* of public forum doctrine for determining both the shape of political discourse and the shape of public space in the city—and hence the shape of ongoing activism in and over public space.

VIOLENCE, ORDER, AND THE
CONTRADICTIONS OF PUBLIC SPACE

Public space is contested both in political and legal theory and on the ground. As a legal entity, a political theory, and a material space, public space (as will be made clear throughout this volume) is produced through a dialectic of inclusion and exclusion, order and disorder, rationality and irrationality, violence and peaceful dissent. In order for material public spaces to "work" as places of politics, the Supreme Court's public forum doctrine declares that they must be orderly and rational. This is an old idea. Public space has long been a place of exclusion, no matter how much democratic ideology would like to argue otherwise (hence the need to go *again* to Hyde Park). Whether in the ancient Greek *agora* (Hartley 1992; Sennett 1994), or the early American republic (Marston 1990; Shklar 1991; Staeheli and Cope 1994), rational "free" discourse was protected by excluding irrationality. In both cases—the *agora* and the American common—women and the majority of men were politically (if not always bodily) excluded. By definition, excluded peoples were irrational and disorderly: rationality was closely defined in terms of gender, class, and race. It resided in the bodies of certain men (E. Wilson 1991).[7]

The ideologies of inclusiveness that support public forum doctrine are at odds with the fact that excluded groups—women, workers, political dissidents, sexual minorities, and all those deemed by dominant society to be disorderly or unruly—have had to fight their way into public if they wanted to be heard (or sometimes even seen).[8] The history of widened representation in public space is thus not one of simple progressive enlightenment by courts or legislatures. To win the right to representation as part of the political public, excluded groups have taken the streets, plazas, and parks and transformed them into spaces *for* representation.[9]

This history of exclusion and struggle thus exposes the contradiction that structures public space. On the one hand, democratic political ideology rests on the assumption that only orderly, rational discourse can produce the sort of "free trade in ideas" (*Abrams v. United States* 1919, 630) that makes truth and informed public opinion possible. On the other hand, order and rationality are preserved by excluding some people and some conduct from the space of the public. Within this contradiction lies the assumption strongly held by the Court that orderly discourse can be preserved in public forums only to the degree that both the spaces and the discourse are devoid of force. For this reason, violent—or even quite forceful—dissent is considered within the law to be fully transgressive of the boundaries of appropriate behavior in public space. Yet, as we will see, often it is only by being "violent" or forceful that excluded groups have gained access to the public spaces of democracy. Indeed, it is precisely such "violence" that has forced the liberalization of public space laws, that has forced what *appears* in law as merely a progressive opening up and more careful regulation of dissent so that more, rather than fewer, viewpoints are heard.

I have put "violence" in quotation marks in the previous sentences simply to raise the question—a question that is frequently before the Court in public forum doctrine cases—as to just what violence is. On the surface, violence appears to be a simple concept: it is the act of doing harm, injury, or desecration through physical force. In social use, however, things get more complicated. As Raymond Williams (1983, 239) has pointed out, it is common to refer to the acts of terrorists as "violent," but not so the acts of their opponents, such as the army, which engage in the use of "force." Moreover, common usage usually conflates the sense of violence as doing harm with violence as "unruly behavior." "It is with the assumption of 'unruly,' " Williams (1983, 239) suggests, "and not, despite the transfer of the word, of physical force, that loud and vehement (or even strong and persistent) verbal criticism has been described as violent."[10] The exclusion of violence from public space has often been simply the exclusion of the "unruly"—those who are a priori defined as illegitimate and thus threatening to the existing order (as so well illustrated by the complaints of "little Arnolds" such as George Will).

Yet it is clear that, despite the arguments of those who seek to protect the status quo in the name of an orderly and limited democracy, violence (from "below") can promote justice or the expansion of rights. This is why the masses went to Hyde Park in the first place, and it is why they must continue to go again and again. As Hannah Arendt (1972, 160–161) has written:

> . . . in private as in public life there are situations in which the very swiftness of a violent act may be the only appropriate remedy. . . . The point is that under certain circumstances violence—acting without argument or speech and without counting the consequences—is the only way to set the scales of justice right again.

The seeming irrationality of violence—"acting without argument or speech and without counting the consequences"—becomes a *rational* means for redressing the *irrationality* of injustice, for withdrawing consent from an order that does not deserve to be legitimated. Violence thus has a (contested) normative aspect.[11]

Although violence has the potential for being just, Arendt (1972) argued that the use of violence by the state usually implies an evacuation of power at the heart of the regime.[12] In her discussion, however, Arendt (1972) purposefully elided the "rational" and "irrational" aspects of violence—violence as impudent harm and violence as a means to an end—showing that no such separation is easy or natural and that violence has different meanings depending upon by whom, on whom, and for what purpose it is committed. Thus, on the one hand, "violence appears as an alternative to institutionalized political influence—the voice of the voiceless, the ultimate, and often effective insistence of the deprived on being taken into account" (Rule 1988, 172). On the other hand, experience of state violence "leads to the expectation that *repression works*" (Rule 1988, 176, emphasis in original). That is, both specific acts of repression and ongoing repressive violence by the state (or dominant factions within the state) make dissent, particularly violent dissent, too costly, leading to conformance within the political system. Only when cracks appear in the façade of unremitting repression can oppositional violence break through the wall of oppression. There are times when dissent *must* be violent. As Tushnet (1984, 1390) has argued, to make excluded voices heard, protests require drama, whether by becoming "the victim of discrediting violence, or developing some novel form of protest. . . . Yet precisely because effective types of protest must be novel, they are not protected in public forum doctrine."[13] Being "unruly" often is a prerequisite for getting heard at all: mere speech is not enough—but it is all that is protected by the Court.

DISORDER, VIOLENCE, AND THE LEGAL CONSTRUCTION OF PUBLIC SPACE BEFORE WORLD WAR I

Picketing as Violence

The Supreme Court's concern with dissent and its relationship to violence in public space is of relatively recent origin. Before World War I, the Court decided no important public speech or assembly cases. When local or lower federal courts addressed the problem of assembly and speech, their concern usually was with exclusion, and in particular with finding ways of stopping demonstrations by militant workers. For most justices, as for most urban reformers or upper-class lawmakers, members of the working class were inherently irrational and violent agents in an urban society that was not just built on, but demanded, a strictly controlled rationality (see American Social History Project 1989, 138; 1992, 93).[14] For many judges, the "typical worker" "was almost always male, single, white, but probably foreign born, with an unpronounce-able last name, uneducated, unskilled, rootless, shiftless, irrational, unpredictable, aggressive, and thus prone to violence, and sympathetic to socialist, revolutionary ideas" (Avery 1988/1989, 7). Judges therefore imagined workers as incapable of properly appreciating "American" institutions and values—such as the right of free speech and assembly—and hence saw workers as threatening to those very rights. Early federal intervention in speech and assembly cases was not concerned either with protecting the right to dissent or with creating the boundaries in which dissent was possible, but rather with controlling the public *behavior* of the working class.

Most typically, this concern over behavior and order can be seen in cases concerning labor picketing. Picketing was not considered speech, or even expressive behavior. Rather, it was an act of violence that had to be tightly controlled or limited. "There is and can be no such thing as peaceful picketing," declared a district judge in 1905 (*Atchison Topeka & Santa Fe Railway v. Gee* 1905, 584), "any more than there can be chaste vulgarity, or peaceful mobbing, or lawful lynching. When men want to converse or persuade, they do not organize a picket line." Under such assumptions employers found no difficulty in obtaining injunctions enjoining all picketing or other means of public demonstration and persuasion. The *assumption* that picketing inevitably led to violence was enough to demonstrate to a court the *necessity* of controlling workers during industrial disputes. The threat of contempt proceedings against union leaders for failing to stop picketing and public displays often was enough to break a strike (Avery 1988/1989; Forbath 1991).[15]

Moreover, picketing had to be banned because it "contradicted the volunteeristic assumptions, individual sovereignty, and respect for property embedded in the law" (Fink 1987, 911). Picketing was, in Justice Holmes's words, a "conspiracy" (*Vegelahn v. Guntner* 1896, 1077), because it demanded group solidarity and constant persuasive pressure on others to join or support the cause. Under common law in the 19th century, union activity was "legitimate only if [it] came within narrow boundaries of socially acceptable behavior and did not threaten the employer, his patrons, or his workers with physical harm," or if it "did not interfere with the employer's freedom of contract" (Avery 1988/1989, 38). If harm (to property particularly) was a form of violence, then all concerted struggle against employers was understood to be violent. The issue for the courts (and for local governments) before World War I, therefore, was how to prevent the violence that was an *inevitable* part of labor disputes. The frequent answer was to restrict the use of the streets by the workers, which not incidentally also served to protect the existing political and economic order (American Social History Project, 1992, 125).

Taking to the Streets

Workers took to the streets anyway, creating "novel forms" of protest, finding that it was only in that manner that their cause could be seen and heard by others in their communities—and that "judge-made law" restricting the rights of protest could be shown as inherently unjust. For such organizations as the Industrial Workers of the World (IWW, or Wobblies), which was founded in 1905 as a radical union and social movement dedicated to forming "the structure of the new society within the shell of the old" (IWW 1990), the streets and parks of American cities were the most important organizing ground.[16] As one IWW organizer declared, "The street corner was [the Wobblies'] only hall, and if denied the right to agitate there, then they must be silent" (quoted in Foner 1965, 172). In the western United States, the IWW concentrated on organizing the poorest of workers—those that the mainstream press derided as transients, hobos, bums, vagrants, and "won't works." Turnover on the job was too great (and repression by local law enforcement and employers too brutal) to allow sustained organizing at the job site (Foner 1965; Dubofsky 1988). To reach these workers, therefore, the Wobblies had to take to the streets of western cities. As importantly, the local mainstream press typically ignored the organizing drives and strikes of the IWW, or reported them entirely to the favor of the employer (Foner 1965; Mitchell 2002). The street was thus the *only* place from which the interest of the IWW could gain currency.

IWW members used the streets well. They were good at drawing crowds, outshouting the Salvation Army as it attempted to proselytize the same migrant or "casual" laborers, and reworking Salvation Army songs to suit the cause of the Union (IWW 1990; Foner 1981). In response, city after city in the West passed ordinances banning street meetings. The ordinances were justified on the basis of clearing the streets to improve traffic flow, or to promote some other "common" good. But given that such ordinances typically were passed during organizing campaigns or bitter strikes, and given that organizations such as the Salvation Army were regularly exempted from the ordinances, it is clear that these ordinances were less concerned with traffic and more concerned with the growing organizing success of the IWW among poor workers.[17]

In response to these restrictive speech and assembly ordinances, the Wobblies engaged in a series of "free speech" battles in western cities between 1909 and 1916. A very bloody but quite typical fight erupted in San Diego in 1912. Citing "congestion," but pressured by sugar magnate John Spreckles and the local Merchants' and Manufacturers' Association, the city council banned all street speaking (except by religious organizations) in a 49-block area downtown where workers typically congregated. With support from allies among the local socialists, AFL members, and the California Free Speech League, the IWW responded by directly defying the ban by holding large street meetings outside the restricted area. Incensed by such resistance to the ban on speaking, the *San Diego Union* (March 4, 1912) advocated "hanging" IWW members and their supporters, reminding San Diego citizens that it was their right and duty to end by any means necessary the "anarchy and disloyalty" that had befallen the city (Weinstock 1912, 17–19). Not surprisingly, those who constituted what an investigator for the state termed "much of the intelligence, the wealth, the conservativism, the enterprise, and also the good citizenship of the community" (Weinstock 1912, 8–9) responded with unbridled vigilante violence (Foner 1965; Dubofsky 1988). Free speech activists were forced to "run the gauntlet" of gun butts, whips, and clubs as they were driven out of town by the "good citizens" of San Diego. Vigilantes tarred and feathered numerous people and savagely beat Emma Goldman's escort Ben Reitman in an action that garnered national press coverage (Bruns 1987; Falk 1990). Ignoring the violent activities of the vigilantes (and often complicit with them), the police violently broke up downtown meetings. They also requested and received from the city council a "move-on" ordinance that gave them the right to disband *any* street meeting within city limits. The police did not hesitate to exercise their new power: "Police clubs were freely used" against workers who sought to speak on the streets the night the "move-on" ordinance went into effect, "and the blood flowed as a result" (*San Diego Union* March 29, 1912). In San Diego, the politics of the street certainly were not guided by the notion that dissent is vital to democracy, the notion that undergirds public forum doctrine, but rather by violence both by the state and by state-sanctioned vigilantes.

Concerned less that rights of speech and assembly were being violently crushed in San Diego and more that vigilante violence was giving California bad press, the governor eventually sent the state attorney general to San Diego to threaten the activation of the military and the arrest of vigilante leaders. The San Diego vigilantes backed down, and the ban on street speaking was repealed—but not before the vigilantes convinced President William Howard Taft to explore the possibility of using federal laws and police power to destroy the IWW altogether (Foner 1965, 202–203). Taft was clearly convinced by reports from the vigilantes that some 10,000 IWWs were conspiring to overthrow the government. Taft therefore asked the Justice Department to explore how the government could "show the strong hand of the United States in a marked way so they shall understand that we are on the job" (Taft to Wickersham, September 7, 1912, quoted in Foner 1965, 203).[18]

Taft's reaction was in accord with decisions he had earlier made as a federal judge. In these, he was concerned with restricting the rights of strikers because their "conspiracies" threatened the interests of private property which it was the state's duty to protect (see, especially, *In re Phelan* 1894; Avery 1988/1989). Taft's presidential concern also presaged his decisions in free speech cases as Chief Justice of the Supreme Court in the 1920s. In 1912, however, the Supreme Court left these matters to be addressed by local jurisdictions and lower courts—and to the use of violence as necessary to control dissent.

MAKING DISSENT SAFE FOR DEMOCRACY

Espionage Cases and a Liberally Repressive Language of Speech

Taft's concerns about IWW conspiracies are apposite for two reasons. First, they were part of a wider governmental concern with radical, presumably subversive, activity (see Preston 1963). This distrust of dissent only grew during World War I, and eventually numerous socialists, labor leaders, and nearly all the important IWW leaders were jailed or deported (or they fled the country to avoid persecution) under repressive espionage and sedition laws passed in 1918 and 1919. Second, a series of cases that came before the Supreme Court in 1919 as a result of concern over antigovernment agitation during World War I required the Court for the first time to develop a language of protected public speech. These cases showed that "conspiracies" (against government or business) and "violence"—unruly behavior—could be divorced from First Amendment guarantees of assembly and speech in a manner that protected formal rights of speech but clearly drew boundaries around how that speech could be conducted and to what uses it could be put. The fear of conspiracy, violence, and subversion that guided the Court decisions in 1919 reinforced a long-standing American distrust of *collective* rights, even as it heightened protections for individual rights (cf. Gostin 1988, 9). By so doing, the Court was able to create speech rights as a "legal fiction" (Halberstam 1993)—as formal guarantees not necessarily supported in practice—that reinforced prevailing arrangements of power, but did not require the use of state violence to shore up that power.[19]

As part of the federal crackdown on dissent during World War I, a quartet of espionage and sedition cases reached the Supreme Court in 1919 (*Abrams v. United States* 1919; *Debs v. United States* 1919; *Frohwerk v. United States* 1919; *Schenck v. United States* 1919). Largely because of the forceful opinions written by Justice Holmes in these cases, today's First Amendment scholars often trace the birth of public forum doctrine to this time. All four cases asked the Justices to decide whether content, location, and intent of public utterances made those utterances illegal. In all four cases the Court held that public utterances could be tightly controlled, indeed—that mere speech could be treasonous because of the "tendencies" inherent in that speech, its location, and its intent.

Schenck: Protecting the State

Charles Schenck, a nationally prominent official of the Socialist Party, had composed a circular encouraging recently drafted men to protest the draft law by petitioning the government. He further claimed that it was the duty of all citizens to support the "right to assert your opposition to the draft" (*Schenck* 1919, 51). For the latter claim, Schenck was convicted under the Espionage Act for attempting to disrupt the draft. Justice Holmes wrote a unanimous decision for the Court upholding the conviction, but in this decision Holmes first adumbrated a set of key phrases that would guide First Amendment speech law until this day. "We admit that in many places and in ordinary times," Holmes wrote (*Schenck* 1919, 52),

> the defendants . . . would have been within their constitutional rights. But the character of the act depends upon the circumstances in which it is done. . . . The most stringent protection of free speech would not protect a man in falsely shouting fire in a theatre and causing a panic. It does not even protect a man from injunction against uttering words that may have all the effect of force.

Beyond the wholly inappropriate analogy linking political speech to causing a panic in a theater (Smolla 1992), Holmes is clearly outlining a doctrine of speech geared toward protecting the interests of the state (Rosenberg 1989). He is far less concerned with the rights of the individual or collective dissenter, as he makes quite clear in his next sentences: "The question in every case is whether the words used are used in such circumstances and are of such a nature as to create a clear and present danger that they will bring about substantive evils that Congress has a right to prevent. It is a question of proximity and degree" (*Schenck* 1919, 52).

The "clear and present danger" test that the Court still uses to presumably protect most forms of public space "was born, thus, as an apology for repression" (Cover 1981, 372)—a repression based on the grounds that, given the circumstances, speech might well prove persuasive. In essence, Holmes's decision merely upheld an older "bad tendency" doctrine that held that people could be punished for making statements "inimical to the public welfare, tending to corrupt public morals, incite crime, or disturb the peace" (*Gitlow v. New York* 1925, 667).[20] The possibility that Schenck *might* persuade others to heed his advice was enough to prove conspiracy and warrant a conviction (Cole 1986, 880–881).

Frohwerk and *Debs*: Conspiracy versus Individual Rights to Speech

Less than 2 weeks after ruling on *Schenck*, the Court decided two more espionage cases, *Frohwerk* and *Debs*. Both unanimous decisions were again written by Holmes. In *Frohwerk*, the editor of a very small circulation German-language newspaper argued that it was foolish for the United States to send troops to France since Germany was too strong, and thus American draftees could not be faulted for refusing to fight. Although he had not advocated violence, or even organized resistance to the draft, and had explicitly condemned antidraft riots in Oklahoma, Frohwerk's conviction was upheld by Holmes on the grounds that the newspaper *might* fall into the hands of drafted soldiers and thereby disrupt the war effort. Even the remotest likelihood that a person's words might prove persuasive was enough for Holmes and the Court to find a conspiracy that trumped any right to free expression. As Smolla (1992, 100) has pointed out, Holmes did not even refer to his "clear and present danger" standard from *Schenck*, "perhaps sensing that *Frohwerk's* trivialities could never be made to fit the language."

Eugene Debs, the most prominent socialist in America, was arrested, convicted, and sentenced to two concurrent 10-year sentences for making a speech on socialism that, in passing, praised the courage of draft resisters who had been arrested and sentenced to lengthy jail terms. In upholding Debs's conviction, Holmes held for a unanimous Court that the key to the *Debs* case was that the defendant *intended* to obstruct the war effort, even if his words did not directly support such a conclusion: "[O]ne of the purposes of the speech, whether incidental or not does not matter, was to oppose not only war in general, but this war, and that the opposition was so expressed that its natural and intended effect would be to obstruct recruiting . . . [and] that would be its probable effect" (*Debs* 1919, 214–215).

Once again the Court held that the mere possibility that speech could be persuasive was enough to show that it *was* persuasive. This persuasiveness, in turn, was enough to show either that a conspiracy existed or that the words would lead to an inevitable and intended result: the obstruction of the war effort. Hence, it was enough to justify the regulation of speech content so that the interests of the state could be protected.

But Holmes's language marked an important distinction that allowed for an eventual liberalization of speech doctrine. In essence, Holmes began to outline a theory of dissent in democracy that hinged upon a distinction between the right of an *individual* to speak (and perhaps even to gain an audience) and the possibility that her or his words might actually have an effect—that they might, through persuasion, lead to conspiracy. Like picketing, persuasive speech was understood to be a forceful, even violent, act. Mere speech, however, so long as it did not possess "bad tendencies," was safe. Public discourse between individuals, each rationally expressing ideas, was democratic speech. Mass speech aimed at persuading an audience was illegitimate (at least if that speech was in dissent).[21]

Abrams: A Liberal Language of Speech Rights

This distinction became explicit in Holmes's dissent from the last of the 1919 cases. *Abrams v. United States* (1919) was decided some 8 months after the others, and Holmes's dissent is seen by many law scholars as a radical break from his earlier decision (cf. Smolla 1992). Holmes himself stressed the continuity of his decisions, and it is just this continuity that in fact *allowed* the Court to develop a repressively liberal language of free speech.

What seems to have worried Holmes during 1919 was that the trio of decisions early in the year did not adequately explore the question of "proximity and degree" he raised in *Schenck*. The *Abrams* case provided an opportunity for further exploration. The case concerned five Russian immigrants who threw leaflets out of a New York window onto the streets below. These leaflets (in Yiddish and Russian) protested U.S. intervention in Russia after the Revolution. The Court, this time with Holmes dissenting, upheld the convictions under the espionage and sedition acts. In his dissent, Holmes argued that the actions of the Russians did not express an immediate enough danger to the interests of the state. To be proscribable, he argued, speech had to be so dangerous that "an immediate check is required to save the country" (*Abrams* 1919, 630). "Clear and present danger" was now more tightly drawn. The *tendency* to cause harm was no longer quite enough for Holmes. Rather, danger had to be immediate.

Holmes's determination on this score comes in the midst of his final dissenting paragraph—"a paragraph that is cited more frequently than many majority holdings" and that "established the theoretical foundation for all subsequent developments of First Amendment law" (Cole 1986, 885):

> Prosecution for the expression of opinions seems to me perfectly logical. If you have no doubt about your premises or your power and want a certain result with all your heart you naturally express your wishes in law and sweep away all opposition. But when men have realized that time has upset many fighting faiths, they may come to believe even more than they believe the very foundations of their own conduct that the ultimate good desired is better reached by free trade in ideas—that the best test of truth is the competition of the market, and that truth is the only ground upon which their wishes can be safely carried out. That at any rate is the theory of our Constitution. It is an experiment, as all life is an experiment. Every year if not every day we have to wager our salvation upon some prophesy based upon imperfect knowledge. While that experiment is part of our system I think therefore we should be eternally vigilant against attempts to check the expression of opinions that we loathe and believe fraught with death, unless they so imminently threaten immediate interference with the lawful and pressing purposes of the law that an immediate check is required to save the country. . . . (*Abrams* 1919, 630)

This is a fully liberal language of speech. Truth is at once tested and revealed through the "free trade in ideas" where privately held beliefs, expressed in public, enter into competition with all other ideas. For any of these ideas to be heard, however, they must be expressed rationally, and they must not threaten the state itself. Rather, the market metaphor suggests that public speech must consist of simple propositions that, like commodities, stand or fall on their own merits. Ideas are independent, even fetishized, entities that gain currency only by their "rightness." Ideas have a force of their own, and so if they are "forced" into the marketplace, they are by definition invalid. But unlike San Diego in 1912, order in public discourse is not defined solely by the exclusion of disorder. Instead it is defined by a need for civility, which itself will assure that all ideas get heard. To allow the market to function "freely," state repression of expressive activity is still permissible in Holmes's formulation as long as the activity can be shown to be threatening to the market itself. Attracting a crowd, attracting adherents, and thus creating a monopoly in the market of ideas could be grounds for state intervention.

In this sense, Holmes's dissent is less a break from his earlier decisions than often assumed: it still rests on the assumption that orderly speech is individualistic, tightly rational discourse. And it says nothing about the relations of power that may govern entrance into the market in the first place. Indeed, Holmes's (and the rest of the Court's) attention was soon diverted from such issues to an examination of how to best promote "civility" in the free trade of ideas. And in so doing, whatever progressive potential that may have existed in Holmes's dissent was soon eclipsed by a desire, in the name of civility, to assure that the free trade in ideas in no way threatened property rights—especially the property right the Court assumed employers had vested in their employees.

Protecting Property through Civility: *American Steel Foundries v. Tri-City Trades Council*

The "disorder" that threatened the civility of the streets and the property of employers in the 1920s was defined not just as the immediate disorder of the streets—although a new spate of free-speech fights certainly indicated the importance of that (Berman 1994)—but also as the threat of global disorder implied by the rise of Bolshevism and fascism in Europe. For conservative jurists and politicians like Taft, the threat of Bolshevism was particularly vexing because it seemed to be behind much of the local disorder in the streets of American cities. Bolshevism "had penetrated this country," Taft warned in 1919 as the Supreme

Court was deciding the espionage and sedition cases outlined above. He continued:

> Because of the presence of hordes of ignorant European foreigners, not citizens . . . with little or no knowledge of our language, with no appreciation of American civilization or American institutions of civil liberty, it has taken strong hold in many of our congested centers and is the backing of a good many of the strikes from which our community is suffering. (*New York Times*, October 31, 1919, quoted in Cover 1981, 353)

At the height of the postwar Red Scare, and in the midst of the notorious Palmer Raids that carted hundreds of activists off to jail, Taft clearly was expressing the tenor of the times. City ordinances throughout the country still forbade public gatherings for political and labor purposes, and Taft saw it as the duty of the judiciary to support the order that these ordinances sought to impose. Taft's opportunity to do just that came when he was named Chief Justice of the Supreme Court in 1921. The first important case he decided addressed precisely the issue of how much speech (or other expressive activity) could be allowed in public space before it threatened the existing order.

American Steel Foundries v. Tri-City Central Trades Council (1921) had started 7 years earlier when the steel company reopened its foundry in Granite City, Illinois, hiring back less than one-quarter of its former workforce—and these at less than the union wage. Despite high local unemployment, the Trades Council called a strike for the plant. When only two rehired workers responded to the call, the Council established pickets around the plant entrances. Three or four groups of up to two dozen picketers were stationed on public streets around the factory. During 2 weeks of picketing, there were occasional outbursts of violence on both sides of the dispute. Citing violence by picketers, American Steel obtained an injunction "perpetually" restraining strikers and union officials from using "threats of personal injury, intimidation, suggestion of danger, or threats of violence of any kind." The injunction also banned the use of "persuasion" by picketers attempting to get workers to join the strike. Similarly, the injunction forbade "assembling, loitering, or congregating about or in proximity" to the plant. In the end, the injunction simply banned all picketing "at or near the premises . . . or on the streets leading to the premises of American Steel" (*American Steel Foundries* 1921, 193–194).[22]

Deciding that picketing "automatically indicated a militant purpose inconsistent with peaceable persuasion" (*American Steel Foundries* 1921, 205), Taft wrote for a unanimous Court that it was the duty of the court system to protect the order of public spaces outside the factory and the private property rights of the factory owners (a right that included access to labor). "If in their attempt at persuasion or communication with those with whom they would enlist" in their cause, Taft argued,

> those on the labor side adopt methods which however lawful in their an- nounced purpose inevitably lead to intimidation and obstruction, then it is in the court's duty . . . to limit what the propagandists do as to *time, manner and place* as shall prevent infractions of the law and violations of the right of the employees and the employer for whom they wish to work. (*American Steel Foundries* 1921, 203–204, emphasis added)

The language here was destined to become quite important in the shap- ing of political activity in public spaces. "Time, place, and manner" re- strictions, like the "clear and present danger" test, are a cornerstone of contemporary free speech and public space law.

For decades, labor had been agitating against the sorts of injunc- tions upheld in *American Steel*, arguing that they all but eliminated the First Amendment rights of labor to "protest and unit[e] peaceably to re- dress wrongs" (Forbath 1991, 141). Against this Taft upheld the lower courts' long tradition of seeing labor protest as an interference with property rights and the freedom of contract of nonunion workers. "This construction," Forbath (1991, 141) has noted, made "legal repression of labor protest unproblematic." Similarly, various courts had no trouble in striking down labor-inspired anti-injunction laws, claiming that these constituted unacceptable "class legislation" and thus were in vio- lation of the Fourteenth Amendment guaranteeing equal protection for all classes under the law (Forbath 1991, 147–158). Yet, this construc- tion also proved problematic—if for no other reason than when courts granted primacy to employers' property rights they too were engaged in a rather transparent form of class-based lawmaking. While Taft and other jurists (including Holmes) did not shy away from simply asserting the primacy of property rights, they often struggled to couch those rights in a universal language that masked the class-based nature of their rulings. This language was typically the Arnoldian language of civility and order.

By definition, and because it was seen to be inevitably violent, pick- eting exceeded the bounds of appropriate "manners" in public space.

The "persistence, importunity, following and dogging" that the Court understood to be always a part of picketing "became an unjustified annoyance and obstruction which is likely soon to savor of intimidation" (*American Steel Foundries* 1921, 204).[23] Municipalities and lower courts, therefore, were required in the name of order to enjoin strikers from picketing. An additional problem with picketing was that it was "certain to attract attention and a congregation of . . . bystanders and thus increase the obstruction as well as the aspect of intimidation which the situation quickly assumes" (*American Steel Foundries* 1921, 204). In other words, *because* it is an effective means of communication, and *therefore* threatened the property rights of the employer, picketing had to be banned. As Avery (1988/1989, 94) points out, the greater the sympathy of a community to the demands of picketers, the more likely that picketing and congregating would be banned.

The right to the city was quite narrowly drawn indeed, as Taft made plain in his suggested remedy in *American Steel Foundries*. Since picketing "cannot be peaceable" (*American Steel Foundries* 1921, 205), Taft sought to replace picketing with what he called "missionaries" stationed at the entrance to the plant. And, Taft reasoned, the behavior of these missionaries had to be carefully controlled:

> We think the strikers and their sympathizers engaged in the economic struggle should be limited to one representative for each point of ingress and egress in the plant or place of business and that all others should be enjoined from congregating or loitering at the plant or in the neighborhood streets by which access is had to the plant, that such representatives should have the right of observation, communication and persuasion, but with special admonition that their communication, arguments and appeals should not be abusive, libelous, or threatening, and that they shall not approach individuals together but singly, and shall not in their single efforts at communication or persuasion obstruct an unwilling listener by importunate following or dogging his steps. (*American Steel Foundries* 1921, 206)

The space of Holmes's "free trade in ideas" was thus governed, according to Taft (and without Holmes's dissent),[24] not by the ability of social groups to get themselves and their ideas represented to a larger public, but rather by a set of restrictions designed to assure that all communication was individual, polite, and nonthreatening to the property rights of factory owners. The Court argued that "courts could, and should . . . prohibit the politics of the street" (Cover 1981, 359), and replace it with a presumably depowered, rational, and orderly discourse.

Even as the Court sought to restrict—or even eliminate—the politics of the street, the issue nonetheless continued to be contested on the streets themselves. While working-class politics and union policy were certainly shaped by decisions such as *American Steel Foundries* (see Forbath 1991), workers continued to picket and to organize in public spaces. And localities continued doing what they had always done when speech and expressive activity seemed to threaten existing political and economic interests: they banned speech and arrested dissenters. "Although cities rarely pursued prosecutions" of violators of speech and assembly ordinances in the 1920s, "they successfully disbanded [street] meetings, thereby suppressing unwanted speech" (Berman 1994, 301). And well into the 1930s, they had the support of the majority of the Supreme Court. As Forbath (1991, 127) concludes, "[b]y legitimating employers' intransigence and the heavy-handed policing of strikes," such ordinances and the case law that supported them "put the onus of violence and 'disorder' upon trade unionists; it meant that even the most respectable trade unionist was always vulnerable to being treated like an outlaw, a 'thug,' or an anonymous revolutionary," and that, by contrast, corporations were trusted as nonviolent and rational entities.

Anti-Picketing Ordinances and the Further Liberalization of First Amendment Doctrine

Given the Court's approval of restrictions on picketing and other forms of labor speech, employer groups in the 1930s turned increasingly to anti-picketing ordinances as tools in their battles with labor. In California, for example, the language of local and statewide anti-picketing initiatives closely mimicked Taft's language in *American Steel Foundries* (LFC 1941, Mitchell 1996c). Attorneys for the promoters of a 1938 statewide anti-picketing initiative cited *American Steel Foundries* to declare that the *only* constitutionally permissible picketing was that done by a single peaceful missionary (LFC 1938).

However, as the muckraking Senate La Follette Committee investigating abuses of workers' rights documented so convincingly, the anti-picketing ordinances in California had the effect of promoting rather than restricting violence. On the one hand, supporters of anti-picketing ordinances saw their enactment as permission to bloodily prevent public assembly and picketing by striking workers. On the other hand, workers understood that the intent of the ordinances was to prevent not just picketing but organization as well, and thus continued to force their way into the streets, parks, and sidewalks whenever strikes or organizing campaigns were in progress. In 1944, the La Follette Committee (LFC 1944 1643) argued that anti-picketing ordinances in California and elsewhere were clearly unconstitutional. But at the time the ordinances were written (1933–1938), Supreme Court precedent was grounded in the decisions of Taft, not in the more liberal interpretations of 1940 that the La Follette Committee cited (*Carlson v. California* 1940; *Thornhill v. Alabama* 1940). These later decisions were hard-won rather than freely given.

As the Depression decade advanced, courts found it hard to avoid the fact that Taft's language from the 1920s was inadequate to the task of regulating or eliminating violence in public space. With militant workers refusing to abide by the dictates of a restructuring capital; with workers only intensifying the "great popular defiance" to judge-made law that had marked the first two decades of the 20th century (Forbath 1991, 141–147); and with a New Deal Congress formalizing workers' rights through the 1935 Wagner Act, the New Deal Supreme Court (with several new Roosevelt appointees) began to rethink its emerging First Amendment doctrine in the hope of better controlling the sorts of violent confrontations that the La Follette Committee was documenting. In 1937, ruling on a criminal syndicalism case from Oregon, the Court for the first time asserted that there did indeed exist a right to assembly that was guaranteed in the First Amendment. The Court ruled that local governments had no right to break up meetings with which they disagreed:

> . . . peaceable assembly for lawful discussion cannot be made a crime. The holding of meetings for peaceable political action cannot be proscribed. Those who assist in the conduct of such meetings cannot be branded as criminals on that score. The question, if the rights of free speech and peaceable assembly are to be preserved, is not as to the auspices under which the meeting is held, but whether their utterances transcend the bounds of freedom of speech which the Constitution protects. (*DeJonge v. Oregon* 1937, 365)[25]

But this is still not a clear affirmation of assembly rights. While meetings could not be banned if they did not *immediately* promote violence

(Holmes's "degree and proximity" argument had been expanded to assert that speech and action were not equivalent), the Court still left the door wide open for banning meetings *if* local governments or police could determine that they would tend toward proscribed speech. As we have seen, there still were plenty of proscribable cases of speech.

Even with this partial affirmation of public assembly rights, workers' rights of protest were not guaranteed. While the 1935 Wagner Act had more or less guaranteed the right to strike, the exercise of that right remained restricted by local anti-picketing ordinances. Mayor Frank Hague, the infamous boss-mayor of Jersey City, New Jersey, saw nothing wrong with repressing the rights of labor so that his city would be more attractive to capital (Walker 1990). His government issued a permit system for demonstrations and picketing and then regularly denied permits to Congress of Industrial Organizations (CIO) organizers and their sympathizers. Rather than inviting arrest by actively defying the mayor (as the IWW would have done), the CIO, working with the American Civil Liberties Union, challenged the legality of Mayor Hague's tactics in court by seeking an injunction against the mayor and his police. At the end of 1938, a district court judge ruled that there was no evidence that the CIO, ACLU, or socialist meetings would inevitably lead to violence (as Hague claimed). The following year the case reached the Supreme Court.

The Court was forced to rule on the validity of using the streets, parks, and public meeting places not just for civilized discourse but for political organizing—for just the sort of persuasion earlier Courts had distrusted. In its strongest defense of public space as a locus for political activity to that point, a plurality (but still not a majority) of the Court held that

> . . . wherever the title of the streets and parks may rest, they have immemorially been used for the purpose of assembly, communicating thought between citizens, and discussing public questions. Such use of the streets and public places has, from ancient times, been part of the privileges, immunities, rights and liberties of citizens. (*Hague v. CIO* 1939, 515)

Many histories of free speech stop at this point in the decision and declare that the Court had finally ratified "a new and hard won right" (Walker 1990, 111). As Walker (1990, 111) has written, "public areas have never been 'held in trust' for discussion of public issues. Repression, by the very techniques used by Mayor Hague, was the grand American tradition." In this regard, *Hague v. CIO* is indeed a landmark case.

But Walker's (1990) analysis ends right where it should begin. Whether or not the streets had been held in trust, they had been *used* by militant groups seeking to represent their cause in public. They had taken the streets and parks, whether such a taking was legally sanctioned or not. All that *Hague v. CIO* did was begin to sanction the process. And as the very next lines of the decision indicate, the reason for this sanctioning was to illuminate a workable language of control.

> The privilege of a citizen in the United States to use the streets and parks for communication of views on national questions may be regulated in the interest of all; it is not absolute, but relative, and must be exercised *in subordination to the general order*, but it must not, in the guise of regulation, be abridged or denied (*Hague v. CIO* 1939, 516, emphasis added)

The issue, then, is the same as it had always been: how can order be maintained in the face of the demand to use public space for organization; and (secondarily) how can rights of speech, deemed necessary for the production of truth, be protected through the imposition of order?

In fact, the Court did not outline (and still has not outlined [Sunstein 1992]) a means to test whether "the general order" was in fact arbitrary, protecting the power of those who could afford to be orderly. By arguing that streets and parks constituted a certain kind of public forum (as the plurality called it in *Hague v. CIO*), the Court had found a means of regulating not speech itself but the *space* in which speech occurred. By switching the focus of attention from speech and behavior to the place in which it occurred, from the speech act to the forum, the Court also paved the way for finally declaring picketing to be protected speech in *Thornhill v. Alabama* (1940) and *Carlson v. California* (1940). It was this move, some years after California's anti-picketing laws were enacted, that allowed the Senate La Follette Committee to argue that the "unconstitutional laws which blanket a large part of the State now deserve the rating of oppressive labor practices. Their continued existence threatens not only decent industrial relations and orderly government, but the very essence of democracy" (LFC 1944, 1644). This is an argument the Committee never could have made, at least in reference to Constitutional law, at the beginning of the Depression. Militant struggle and constant defiance on the streets and in the courts by those who had long been defined a priori as disorderly and violent had forced the Court to find a new way to write the rules for civility in American public space—to make dissent safe for democracy.

REGULATING PUBLIC FORUMS

After *Hague v. CIO* it was no longer the goal of courts to proscribe discourse in order to protect the state or property so much as it was to regulate the public forum toward these ends. For Justice William Brennan in the 1960s, this change in focus implied a change in metaphor (Cole 1986). Holmes's "free trade in ideas" no longer quite captured the issue. Now, in Brennan's words, concern was with the market*place* of ideas. Writing in a case concerning whether the Post Office had the right to detain unsealed foreign "communist propaganda," Brennan argued that the right to speak also implied a right to listen, to receive messages, to *be* persuaded. "The dissemination of ideas can accomplish nothing," Brennan wrote, "if the otherwise willing addressees are not free to receive them. It would be a barren marketplace of ideas that had only sellers and no buyers" (*Lamont v. Postmaster General* 1965, 398). Eight years later, Brennan made his concerns even clearer in vigorous dissent in a case concerning political speech on television: "Freedom of speech does not exist in the abstract. On the contrary, the right to speak can only flourish if it is allowed to operate in an effective forum—whether it be a public park, a schoolroom, a town meeting hall, a soapbox, or a radio and television frequency" (*CBS v. DNC* 1973, 193).

To a large degree, the Court has now accepted Brennan's argument (though it has not done much to open the airways to dissenting voices), and the recent abortion cases have been understood by both sides of the Court as concerned with place regulation—with the nature of the forum itself. As David Cole (1986, 891) has argued, the marketplace of ideas metaphor, and by extension public forum doctrine, "ultimately justifies affirmative intervention by the government in order to save not the state, but the marketplace itself." The focus therefore is on the nature of space and its role in promoting or denying the "free trade in ideas."

But both economic metaphors (Holmes's free trade in ideas and Brennan's marketplace of ideas) are quintessentially liberal formulations and thus write power out of the equation by assuming that all actors have equal access to the market. All ideas, all actors, exist as commodities, ready to be bought and sold, always freely circulating according to the logic and dictates of the market except when the state wrongly intervenes to distort the market process. In other words, state intervention is acceptable if it *protects* the market (even if it is deeply iniquitous), but it is unacceptable if it disturbs the market (even if the goal is to *decrease* iniquitous relations and therefore *increase* participation). In this sense, public forum doctrine concerns itself with negative rights (see Chapter 1, note 18)—protection from state interference—and not explicitly with positive rights—the right *to* speak. Hence, public forum doctrine is not so much an assurance that marginalized groups can be heard (the better to promote democratic society) as it is a theory of laissez-faire government, which, given the concern the Court has long expressed about "class-based" legislation, is hardly surprising.

Justice Brennan recognized the degree to which his marketplace metaphor ignored the power relations extant in society, and thus reinforced inequalities (but by doing so he remained in a minority in his own Court—a minority that has since only grown weaker). He argued that those who can communicate through ordinary channels of modern discourse—because they either own and control the media or can readily buy access to it—have little need for popular demonstrations in the street. Brennan was especially concerned that the Court had failed to treat television as the public forum it should have been, writing about the actions of a Court he disagreed with:

> Thus, as the system now operates, any person wishing to market a particular brand of beer, soap, toothpaste, or deodorant has direct, personal and instantaneous access to the electronic media. He can present his own message, in his own words, in any format he selects, and at the time of his own choosing.[26] Yet a similar individual seeking to discuss war, peace, pollution, or the suffering of the poor is denied the right to speak. (CBS v. DNC 1971, 2000)

The legal language of public space may have changed since William Howard Taft's day from the formal denial of rights to their formal protection, but the effect is pretty much the same. The Court remains an instrument—through the language of law—of assuring exclusion in the name of social order.

In this sense, for labor and other dissenting groups, the switch in metaphorical focus has been a Pyrrhic victory, because while it is the responsibility of the state to order space in the "interests of all," the Court has been unable to indicate how the interest of all is to be determined (see *United States v. Kokinda* 1990). The law treats all equally. But all are not equal, and such equal treatment simply serves to reinforce unjust social relations, as scholars such as Mark Tushnet fear (see Chapter 1). The law has no way to recognize that, in order to be represented in public, dissident groups have had to make their claims in a manner that does not conform to constrictive norms and practices of rational discourse—that the needs of those who wish to use public space as a public forum may not at all align with the images the Court holds of orderly, rational discourse. Picketing (for example) works for marginalized groups because it demands notice in a way that dispassionate discourse simply cannot. Orderliness can thus quite easily serve power, as Taft well recognized. The guarantee of the *right* to speak in public forums is quite different from the question of effective access to that forum by those who *need* to speak in the street. For this reason, then, Tushnet (1984, 1387) argues that First Amendment law "has replaced the due process clause as the primary guarantor of the privileged. Indeed, it protects the privileged more perniciously than the due process clause ever did."

Tushnet is surely right in many ways. But he is not completely right, for, as with the Court as a whole, such a claim can only be made if analysis of First Amendment debates and lawmaking are abstracted out of the actual political struggles from which they necessarily emerge. Such abstracting is exactly what the Court does. The *terms* of that abstraction, however, are determined at least in part in public space itself—in the real struggles to speak and be heard, to listen and to gather, to protest and picket, that continually shape the right to the city, either in concordance with or in defiance of the ways that legislative bodies and courts seek to regulate public space and the activities that take place in it. Neither law nor public space is neutral or immutable. Both, in fact, are sources of power, available to be used by those best able to "capture" them and turn them toward their own particular interests.

When law and space come together, as they inevitably do, each structuring the other, it makes little sense to abdicate the language of rights (as many progressives such as Tushnet are willing to do), for this runs the risk also of abdicating not just the language of justice but its practice too. "Rights have not gone away," Blomley (1994b, 410) correctly argues. They remain a key rallying cry, especially among the dispossessed or marginalized. "As such, the dismissal of rights-based struggle as incoherent or counterprogressive seems condescending" (Blomley 1994b, 410). Moreover, it concedes the language of rights to those who offer a discourse of rights "centered on negative liberty, property and the individual" (Blomley 1994b, 412). But since the Court is more a friend of negative rights and the interests of the powerful than of positive rights and the interests of the marginalized, it remains essential that activists move again and again into the streets—that they "return again to Hyde Park"—where a more positive vision of a just society can be fought for (cf. Blomley 1994b, 413), for it is only there that *geography* can be reconfigured in such a way so that law has to pay attention. For women, African Americans, and all manner of ethnic groups, workers, and progressive activists, the fight to claim the streets, parks, courthouses, and other public spaces of the city is precisely the fight to reclaim their rights as members of the polity, as citizens who have both the duty and the right to reshape social, economic, and political life after an image perhaps quite different from the laissez-faire liberalism promoted by the Supreme Court.

CONCLUSION

That is why the 1994 *Madsen* and 2000 *Hill* abortion clinic decisions make a troubling lens through which to focus legal geographies of public space, especially if one of our goals is to elucidate what a *progressive* right to the city might be. In both cases, anti-abortion activists successfully claimed the language of free speech and public protest not to fight for the expansion of rights, and certainly not to hold the state or capital accountable for repressive practices, but precisely to *repress* others' rights: the right to safe abortion, the right to work, the right to enjoy a peaceful home life.[27] In its turn, the Supreme Court both drew on and honed its public forum doctrine so as to better order and regulate not violent anti-abortion activists but protest itself. That is to say, by claiming *not* to look at the *content* of the anti-abortionists message, the Court developed a means by which its regulatory powers could be extended into other protests, other social movements: law derived from a specific struggle could be universalized. In the process, the Court did quite little indeed to advance the positive right of women to abortion: clinics are still few and far between, doctors still ill-trained, levels of violence against abortion providers still on the rise, and women entering clinics still subject to a barrage of "counseling," even if now at a greater distance than before. Anti-abortion activists did not lose much—and certainly not as much as Justice Scalia claims—in these decisions.

It could be argued, in fact, that anti-abortion activists were guaranteed something of a victory (again despite Scalia's dire warnings) simply because the Court chose to see the case as one concerning state intervention into *speech* rather than one concerning a woman's right to abortion. As cases concerned with regulating the public forum, with determining just how to assure minimal state intervention into speech while guaranteeing order, both *Madsen* and *Hill* at once confirm the history of public forum doctrine—a doctrine designed to protect rights in the abstract rather than address the relations of power that make those rights necessary—and further solidify the role of the state in promoting a certain vision of order in democracy. As cases about free speech, *Madsen* and *Hill* will prove troubling to those, like workers engaged in battles with employers, who now more than ever need to take to the streets to press their claims through picketing and rallying. Such a vision of order, as we will see in later chapters, is also deeply troubling for the homeless who must assemble and sometimes even "speak" in the public spaces of the city just to survive. Had these cases been decided in terms of protecting a woman's right to abortion, however, the cases would prove far less troubling to those, like the homeless and workers, who must always seek spaces for representing their demands for the expansion rather than the denial of rights as a means of solidifying "the 'best' aspirations of a people for the society"—a potential for law that Blomley and Clark (1990, 435) argued can be made every bit as "plausible" as the coercive aspects of law that the left is so adept at recognizing. Matthew Arnold and all his little followers, in other words, could be turned on their heads and law could be made a means for achieving, rather than thwarting, the right to the city.

Justice Scalia, in a departure from his earlier work on the Court (see Brisbin 1993), had sought in his dissents to both *Madsen* and *Hill* to make a strong defense of the right to protest *despite* the potential violence that such a right necessarily contains. He argued in *Madsen* that those who were not "friends of liberty" would seize the powerful "loaded weapon" the Court had left lying around and destroy precariously held rights.[28] And so they did, but not at all in the way that Scalia had predicted. Rather, less than a month after the *Madsen* decision, anti-abortion activist Paul Hill grabbed a loaded gun and murdered Dr. John Britton and his escort, James Barrett, outside a Melbourne, Florida, women's health clinic. This was the second murder in 18 months of a Florida doctor who had performed abortions. Paul Hill was tried and convicted under a new federal "access to clinics" law that makes criminal activity around an abortion clinic a federal crime. Even so, many prochoice activists have called for stronger regulation of the public forum. They argue, like the Court before them, that such regulation will not only save lives (because other criminal statutes are presumably not enough) but also will allow for a more reasoned debate about abortion. The Colorado law at stake in the *Hill* case, and similar laws across the United States, is the result of such reasoning. But whatever the efficacy of these laws in the immediate vicinity of clinics, they have done little to stop the murder of abortion providers, as the assassination of Dr. Barnett Slepian in his suburban Buffalo home in October 1998 (most likely by James C. Kopp, who as of this writing is standing trial for the crime) makes clear.

It seems to me that the strategy of regulating space so as to order debate—as access to clinic laws attempt to do—falls into exactly the trap that snared the Supreme Court. It presumes, first, that ideas and actions are immediately separable (that there can be, for example, a "rational" opposition to, or promotion of, abortion that is not linked to the actions of the holders of these views), and that one can be protected at the expense of the other. Second, it presumes that the regulation of certain kinds of (otherwise legal) conduct will protect speech and other rights. If the historical geography of the public forum—and public space more generally—shows anything, it is that this is a fallacious assumption, one that assumes a narrow definition of order is in the interests of all. And since it is not, it will always call up, as we will see in the next chapter, its own opposition. I am not certain that this is a direction that progressive activists—those most concerned with struggling for a right to the city and those who so often have to take to the streets themselves in order to be heard—want to turn. Rather, we need to find ways to enhance *positive* rights without at the same time increasing the state's ability to circumscribe negative rights. The murders of Drs. John Britten and Barnett Slepian have shown that the regulation of space in the name of order will do little to detour those who seek to deny the right to abortion by violent means. Nor has such regulation done much to slow the trampling of labor rights. Indeed, the history of labor and its use of public space shows, on the contrary, that such regulation might do much to help those who would undermine the rights of others.

To explore this point more fully, it is useful to turn away from labor and abortion to a specific struggle for the right to gather and speak in public space: the 1964 Free Speech Movement and the subsequent turmoil in Berkeley, California. In the struggles in and for Berkeley during the 1960s we can glimpse just how important *space* is for any decent right to the city. We will also see that no matter how hard various agents of the state—including the courts—work to make dissent safe for democracy, a vigorous democracy must ever be one in which dissent exceeds the bounds placed on it, one in which people do not just go again to Hyde Park but rather actively *take* Hyde Park and make it into something altogether new.

NOTES

1. To my knowledge, this opinion has not yet been applied to cases concerning panhandling (see Chapters 5 and 6). It does seem, however, to provide the perfect language for upholding the sort of anti-homeless bubble laws that have become prominent in recent years in American cities. These laws prohibit panhandling around automatic teller machines, around doorways to businesses, or around cars. Many aggressive panhandling laws specifically outlaw following, "dogging," or "approaching" people who do not want to be panhandled, just as does the law adjudicated in *Hill*. A review of the relationship between urban geography and free speech jurisprudence as it relates to panhandling can be found in Mitchell (1998a).

2. Colorado's statute, FACE, and other similar laws address only a small part of the problem. Women's effective right to abortion is blocked in numerous other ways, ranging from a shortage of abortion clinics (and their uneven geographic distribution, which means some women have to travel hundreds of miles to receive an abortion) to a rapidly declining number of doctors trained in the procedures, to laws restricting certain surgical procedures, to parental notification laws that deter under-18-year-old women from seeking abortions, to laws restricting the use of public funds to pay for abortions or abortion counseling. One of the key themes of this volume is that a right is only a right to the degree that it is *practiced* as such: paper rights are no rights at all unless people can—or struggle to—engage in the behavior that is putatively protected. Encoding rights in law is important and necessary, but it is not anywhere near sufficient. No clearer example of this maxim can be found than the case of abortion in the United States.

3. One can still support the majority's desire to find ways to protect and promote a woman's right to abortion while admitting that Justice Antonin

Scalia is in fact correct when writing in dissent that the Colorado statute restricts a person's activities based on *what he intends to say* when he" approaches a woman entering a health clinic (*Hill* 2000, 2503, emphasis in original). As we will see, this is precisely why "time, place, and manner" restrictions on the right to speech and assembly are so interesting, so important, and so frequently violated.

4. The interesting thing about both these cases is that, while the Court was sharply divided, both sides—the more liberal branch seeking to regulate protest around clinics so that women would be relatively free from interference from protesters and the more conservative side hoping to discourage the right to abortion—drew on public forum doctrine to make their case. The conservatives, headed by Antonin Scalia, argue that both cases "like the rest of our abortion jurisprudence . . . is in stark contradiction to the constitutional principles we apply in other contexts" (*Hill* 2000, 2503). That is, Scalia argues that the Court regularly sets aside public forum doctrine when it suits their purpose. He is certainly right, as we will see, but the issues are far more complex than his polemic lets on.

5. In some instances private property can assume the function of a dedicated public space, though the owners of this property typically retain an even greater proprietary right to shut out the public—or to regulate even the content of speech.

6. Scalia is absolutely right when he warns in *Madsen* (1994, 2508) that the "labor movement, in particular, has good cause for alarm" at the Court's decision, though his own history is rather truncated and does not acknowledge that the labor movement, in particular, has almost *always* had good cause for alarm since so much First Amendment jurisprudence has developed precisely as a means of neutralizing workers' dissent to capitalism.

7. Current examinations of the ways that public space reinforces normative heterosexuality reveal that such exclusions of normative "irrationality" have not been left behind. For an important and insightful account of the relationship between "the closet" and the public and private spaces through which it is materialized, see Brown (2000).

8. The politics of visibility is always complex. While it is true that publicity has been a critical factor in the extension of women's rights and gay rights (to name just two), so too has it often been the case that *privacy* is a precondition of the development of gender and sexual identities (see Chauncy 1994; Hubbard 1998, 2001).

9. This formulation, and its roots in Lefebvre's arguments about the production of space, will be explored in greater detail in Chapter 4.

10. Obviously this line between "unruly" and "violent" has become even more significant following the wake of the terrorist attacks on the World Trade Center and the Pentagon. In their wake, anticorporate globalization activists immediately began debating whether or not to continue protesting against such institutions as the World Bank and the International Monetary Fund for fear that their street theater (and the occasional small-scale

violence that accompanied it) would draw a disproportionately brutal response from the state. Such a fear was, of course, not ungrounded, given the killing of a protester by police in Genoa in the summer of 2001 and the widespread indiscriminate arrests in Quebec City in April of that same year (including the arrest, on weapons charges, of activists who catapulted stuffed animals across the fence constructed to keep delegates to the Free Trade Area of the Americas summit well out of sight).

11. This is the reason, also, that much Gramscian-inspired work is off the mark when it sets up "consent" and "coercion" as opposites (where one is resorted to when the other fails) rather than always dialectically entwined and in fact inseparable. Gramsci (1971, 12) himself contributed to this error by arguing that "state coercive power . . . 'legally' enforces discipline on those groups who do not 'consent' either actively or passively." This is true, but it is also a special case. Both violence and discipline are ubiquitous (as Foucault so persuasively showed).

12. This, of course, is also one of the main implications of Gramscian notions of hegemony (Gramsci 1971; see also Williams 1977).

13. As we will see, these novel forms are often *formative* of public forum doctrine for the simple reason that judge and legislators as regulators of public order—and legitimators of the state—must *respond* to these protests and find ways to account for them and to bring them into the fold of the state itself.

14. Mona Domosh's (1998) analysis of the "polite politics" that ruled New York public spaces in the second half of the 19th century confirms this point in its analysis of the small, often quite invisible, ways in which upper-class women and men "resisted" their domination in public space.

15. Injunction contempt proceedings do not require a trial by jury, adherence to standard rules of evidence, nor even a showing that violence or other sanctionable activities had occurred.

16. The IWW specifically attacked injunctions controlling workers' protest behavior, arguing that the courts were simply puppets of the capitalist class. Thus, part of its official policy declared: "Strikers are to disobey and treat with contempt all judicial injunctions" (see *Industrial Relations* 1916, Vol. 11, 10578).

17. For a case study of one such controversy (in Denver in 1913) that places the legal struggle over and the production of public space within the context of a complex politics of scale related to regional development, see Mitchell (2002).

18. The IWW was eventually suppressed and all but destroyed by the federal government—with a great deal of help from California state agencies (including one run by the nephew of Harris Weinstock, the San Diego investigator)—during World War I. See Dubofsky (1988); Foner (1965); Mitchell (1996a); Preston (1963).

19. In this move we can see the roots of the distrust of rights that animates such commentators as Tushnet and Rorty, as discussed in Chapter 1.

20. The bad tendency doctrine derives from English common law of libel. For

two examples of Holmes's use of the bad tendency test before Schenck, see *Patterson v. Colorado* (1907) and *Fox v. Washington* (1915).

21. Even as free speech doctrine has been liberalized over the course of the past century (see below), this particular theory has, if anything, been expanded. As courts have sought to "balance" the right to speak with a set of other state concerns, the result has been a spatial regime of regulation in which people can say just about whatever they want, just so long as it is never heard in such a way as to make a difference. In the United States, the right to speech does not seem to entail a right to be heard. I analyze this dynamic in Mitchell (2013a).

22. The family resemblance between the specific geography of protest and control in this case and that regulated by contemporary "bubble laws" is clear.

23. Mark these words. They were rather closely echoed in a series of cases in the last decades of the twentieth century seeking to criminalize the behavior of homeless people (Mitchell 1998a; see Chapters 5 and 6).

24. A week later Taft lost both Holmes's support and that of Louis Brandeis (the other great liberal Justice to whom much public forum doctrine is traced) in a similar case (*Truax v. Carrigan* 1921). The dividing issue here was not whether the politics of the street could be prohibited but rather whether the federal judiciary should allow states to experiment more widely in loosening strictures on street politics if such experimentation might lead to a greater risk of disorder (see Cover 1981, 361–363).

25. Significantly, *DeJonge* was a labor case. Dirk DeJonge was arrested for criminal syndicalism (an old law meant to wipe out the Wobblies) when he helped organize a Communist Party-sponsored meeting protesting police shootings and raids on homes of striking longshore workers.

26. And, one might add, with the lavish attention that the media as a whole pays to various advertising campaigns and events (such as the annual media frenzy over Super Bowl advertising), an ad campaign can now expect to be *amplified* across the range of media, assuring that even those who, for example, do not watch TV are fully apprised of what is being advertised, when, and how.

27. It should be noted that the strongest restrictions on speech upheld in *Madsen* were around private residences and hinged on questions of private property and not speech or assembly per se.

28. In *Hill* he argued that it was the Supreme Court itself that had grasped this "loaded weapon."

3

From Free Speech
to People's Park

*Locational Conflict and the Right
to the City*

How much farther do we have to go to realize this is not just
another panty raid?
—GOVERNOR RONALD REAGAN (May 20, 1969)

Conflict over rights often resolves itself into conflict over geography, as
the Supreme Court's evolution of public forum doctrine has made plain.
Space, place, and location are not just the stage upon which rights are
contested, but are actively produced by—and in turn serve to structure—struggles over rights. Conflict over rights can therefore be understood, at least in part, as a species of *locational conflict*.[1] Rights have to
be exercised *somewhere*, and sometimes that "where" has itself to be actively produced by taking, by wresting, some space and transforming
both its meaning and its use—by *producing* a space in which rights can
exist and be exercised. In a class-based society, locational conflict can be
understood to be conflict over the legitimacy of various uses of space,
and thus of various strategies for asserting rights, by those who have
been disenfranchised by the workings of property or other "objective"
social processes by which specific activities are assigned a location. In
this sense, locational conflict is often *symbolic* conflict, in that the conflict is waged through the deployment of highly symbolic actions. That

81

is, it is waged through a combination of speech and action—the two things the Supreme Court works so hard to keep apart. In fact, the very space of struggle itself comes into being and is defined in locational conflict *because* speech (communication) and action (conduct) are simply inseparable. Further, and again because speech and action are inseparable, geography matters.

That might seem axiomatic, or in fact just tautological—that in locational conflict geography matters—but it is surprising how often it is forgotten that in any kind of social struggle, even struggles regarding place and location, geography, or more precisely the ongoing *history* of locational conflict, is simply forgotten. Take, for example, a recent article in *The Chronicle of Higher Education* detailing what the paper sees as a new trend in speech codes: the development of specific "free speech zones" on college campuses (Street 2001).[2] More and more campuses, according to the *Chronicle*, are developing specific places in which free speech is allowed and restricting it in others as a means of balancing "between universities cherishing the right to free speech and needing to run an institution," as a Dean of Students from UC Berkeley puts it (Street 2001, A38). The *Chronicle* argues that the development of such zones continues a history of debate over speech codes that erupted in the 1980s when several universities attempted to regulate hate and other harassing speech. Many of these codes were struck down by the Supreme Court, and universities thus turned to public forum doctrine to assert their legitimate right to regulate the "time, place, and manner" of speech. There is nothing particularly wrong with this history until the paper asserts that "Tufts University may have been the first [to create a free speech zone]. In 1989, the university, in an attempt to restrict so-called hate speech, designated 'free speech zones' in certain areas of the campus" but quickly dropped the policy when students protested (Street 2001, A30).

The problem with this account is that, despite the fact that the 1964 Berkeley Free Speech Movement (FSM) is referenced in several places in the article, the *spatial* history of that movement—the very fact that the movement erupted in part as a result of the university's attempting to create and enforce specific free speech zones, what the university called "Hyde Park areas"—is lost. That history, as we will see, not only was concerned with the right to speak but also developed as a struggle for an appropriate *place* to speak. Lost too in the Chronicles account is the fact that nearly all California public universities quickly developed specific free speech zones—often in heavily trafficked locales—in response to the Berkeley FSM. Tufts was not first university to demarcate a free speech zone on campus, though it may be the case that the specific politics of regulation driving the current wave of zone demarcation is different than it was in the 1950s and 1960s.[3]

Exploring one of these earlier attempts to zone speech, and the famous reaction it called up, the Free Speech Movement, will help us see that by examining conflict over speech as conflict over location we can learn a great deal about how rights are fought for, claimed, undermined, and reinforced in "actually existing capitalism." Let us delve, therefore, into the specific spatial history of the Berkeley Free Speech Movement in particular, and the changing radical politics of Berkeley in the 1960s more generally. Doing so will shed a good deal of light on current attempts to zone speech and conduct, attempts often couched not as a means of eliminating dissent but of promoting "quality of life."

NONCONFORMISTS, ANARCHISTS, AND COMMUNISTS: FREE SPEECH IN BERKELEY

As a semipublic property, as something like a "dedicated public space" (in the language of the Supreme Court's public forum doctrine), the UC Berkeley campus became an early staging ground in the battles over the redefinition of political, property, and social rights that wracked Berkeley (and the nation) in the 1960s. Clark Kerr, the president of the University of California from 1958 to 1966 (when he was removed from office in one of Governor Ronald Reagan's first official acts), understood what was at stake in the first militant battles over free speech at Berkeley in 1964:

> A few of the "non-conformists" have another kind of revolt [than one against the university] in mind. They seek instead to turn the university, on the Latin American or Japanese models, into fortresses, from which they can sally forth with impunity to make their attacks on society. (quoted in Draper 1966, 206)

For his part, Kerr had a rather different vision for the university in modern society.[4] Writing in *The Uses of the university*, Kerr (2001 [1963]) saw the university and surrounding community as being, in part, a laboratory for the creation of a new and more rational society. The university had an important role to play in the drive toward a rational and managerial political economy. Relabeled by Kerr, the "multiversity," the university was to specialize in the "production, distribution and consumption of 'knowledge' " even as the surrounding city was to be reconfigured to more efficiently reproduce the "workers" who were to perform this production, distribution, and, to a large extent, consumption of knowledge.[5]

Kerr's vision, however, extended well beyond the university and its immediate neighborhood. He was just as keen to describe the new society that was coming to fruition at mid-century. In this new society, Kerr wrote in *Industrialism and Industrial Man* (Kerr et al. 1960), politics too would be made rational or, more accurately, managerial. Men and women "can be given some influence" in the new society, Kerr intoned.

> Society has achieved consensus and it is perhaps less necessary for Big Brother to exercise political control. Nor in this Brave New World need genetic and chemical means be employed to avoid revolt. There will not be any revolt anyway, except little bureaucratic revolts that can be handled piecemeal. (Kerr et al. 1960, 295).[6]

Such pronouncements—which seem to accord rather well with the political pessimism of the later postmodern, post-structuralist left—at the time drew immediate fire from around the globe. Guy Debord (1994 [1967], 137–138), for example, attacked Kerr directly in his 1967 manifesto, *The Society of the Spectacle*, asserting that Kerr's vision was exactly what had to be fought against if people were ever to regain control over their own alienated lives and learn once again to *live* in the city.

Closer to home, Kerr's vision was enacted in part through the University of California's attempts, beginning in the early 1950s, to gain control of the South Campus area (centered around Telegraph Avenue), both for campus expansion and to better control the mix of residential and business functions. A 1952 Long-Range Plan proposed that the university expand into the South Campus area as part of a large city-wide redevelopment program that was aimed at addressing the "blighted" sections of the city. Students and the elderly who lived there were not expected to mount a particularly effective opposition to the purportedly benign plans of the university and the city. As the journalist Robert Scheer (1969, 43) later contended, the bureaucratic motives of the administration were

. . . based on assumptions about the purpose of the University and the role of its students. South Campus expansion was based on the presumed need to sanitize and control the University environment. The university community which the Development Plan envisioned was one of a total environment in which every need—classrooms, housing, recreation and parking—was programmed for ten years into the future. Students would literally be forced to dwell within an ivory tower of concrete and glass dormitories which—along with other official buildings, churches and a few spanking new store fronts properly up to code—would be the only structures permitted in the central South Campus area. All others would be pushed out by the University Regents exercising their power of eminent domain. This would, as the Development Plan (1956 revision) noted, provide "a well-rounded life for students. . . . " If the Multiversity was to be a knowledge factory, South Campus would be its company town.

Just this vision of the university and city as a rational technical and efficient future, carefully managed by competent and well-trained bureaucrats working in the interest of society, became the focus of revolt and popular rebellion in Berkeley in the 1960s rebellion for which the Free Speech Movement is often presented as the opening act.

But the FSM was not simply a spontaneous, massive, inexplicable act of refusal (as many histories have it). Instead, the FSM which shook the Berkeley campus during the fall of 1964 was a climax of a growing—actually rejuvenated—and ever more militant movement against the dictates of a class- and race-based society that refused to grant blacks, workers, and students those rights that were supposedly the very foundation of its existence. By 1964, Berkeley already had a long history of student activism. The 1930s, for example, saw significant student organizing, often led by Communist Party members and their allies, in support of striking farmworkers, longshore workers, and other militant unionists around the state. So too were many students (and faculty) involved in broader "popular front" organizing. In the 1950s the loyalty oath controversies had seen significant student support for resistant and fired faculty. By 1957 a radical student party, SLATE, had formed. And Berkeley students, like their counterparts in many other northern universities, were involved with civil rights struggles, labor struggles, anti-McCarthy actions, and fledgling new-left organizations such as the Students for a Democratic Society throughout the late 1950s and early 1960s.[7]

Be that as it may, proximate causes were important. FSM was in part a clear revolt against the increasingly restrictive policies of a campus administration, directed by Clark Kerr as president of the whole university system, that viewed itself as a center of liberal (capitalist) intellectualism. The American public university campus—and the Berkeley campus in particular—had always been a tightly controlled space. In spite of the history of free speech struggles in the first two decades of the 20th century that forced a reconsideration of laws governing public space, the public universities of California continued, as late as 1964, to operate as if restrictions on the political activities of their students both on and off campus were not only their right but also their mandate. Somewhat unusually among large public universities, the University of California retained the belief that paternalistic *in loco parentis* was a viable and necessary ideology of social control over students.[8] As Columbia University Professor Robert Paul Wolff (1966: 38) wrote in response to an angry article critical of FSM by former Berkeley Professor Lewis Feuer (1966): "In a morally sound society, the university can and should be a sanctuary of scholarship, a school for citizenship, and a validator of the dominant values of the political community." Through a series of rules and regulations designed to severely proscribe what could be said on campus—and where it could be said—this was exactly what the University of California was attempting to do. Among the many issues at stake in the FSM at Berkeley was the question of what was moral and who had the *right* to determine that morality. But, even so, the movement resolved itself, quite explicitly, into a question of the right *to space*. Free speech at Berkeley, as with free speech anywhere, was a spatial problem.

The Geography of Free Speech 1: Context

The Berkeley campus in the 1960s was growing rapidly. The traditional edge of the campus was Sather Gate on Telegraph Avenue (Figures 3.1 and 3.2). In 1960 and 1961 new campus buildings, housing the bookstore, student union, student government, restaurants and coffeehouses, were opened just outside Sather Gate. Telegraph Avenue was closed at Bancroft Way, and the former street was converted into a large plaza. Overlooking the plaza—indeed, dominating it—and also outside the Gate was the main building of the system-wide administration, Sproul Hall (Figure 3.3). Sproul Hall had been deliberately built outside the Berkeley campus in 1940 to symbolize the independence of the campus administration housed on the Berkeley campus proper from the university-wide administration now housed off-campus; the 1960 expansion of the campus, therefore, incorporated the system-wide admin-

FIGURE 3.1. An aerial view of the UC Berkeley, 1965. The campus expanded rapidly southward (to the right in this picture) in the postwar period, stretching beyond Sather Gate (# 4). The complex of buildings labeled #1 are the student union buildings built in the early 1960s. #2 is Sproul Hall, now part of the Berkeley campus; Sproul Plaza is #5. The plaza the Berkeley administration wanted to designate as the "Hyde Park area" is below #1 in this picture. Photograph originally published in Heirich (1971).

FIGURE 3.2. Sather Gate, the traditional entrance to the Berkeley campus. The view is to the south from inside the "old" campus toward the new developments of the 1950s and 1960s. The building in the background is the Associated Students center. Sproul Hall is out of the picture to the left; the plaza through the gate and before the student center is Sproul Plaza. Photograph by author.

istration back into the campus itself. The land upon which the plaza was built was ceded to the university by the city at the time of the street closure. Additionally, the university was engaged in an aggressive program of building student dorms off-campus several blocks south of Sather Gate, in the center of the "blighted" South Campus area (Heirich 1971; Scheer 1969).

All this detail is important because the city street in front of Sproul Hall had for a long time been a traditional *off-campus* free speech area. Student and community activists had long used it as a rallying ground. Indeed, it was the most important political forum in the city. But now it had been incorporated into the campus itself and was thus not subject to the regulations of a "traditional public forum"; instead, the rather more restrictive rules allowed a "dedicated public forum" obtained. Not that the university was much concerned with the niceties of public fo-

FIGURE 3.3. Sproul Hall. Long the home of the university-wide administration, Sproul Hall had been built outside Sather Gate to help reinforce the Berkeley campus's relative autonomy vis-à-vis the administration of the university as a whole. Campus development during the 1950s and 1960s engulfed Sproul Hall, and Telegraph Avenue in front of it was closed to create Sproul Plaza. The steps of the hall and the plaza are the *locus classicus* of the Free Speech Movement and remain to this day the central site for political activity on the Berkeley campus. The university administration moved several blocks off-campus in the 1970s and further decamped to Oakland in the 1980s. Photograph by author.

rum law: it had no qualms about regulating either particular activities (conduct), such as soliciting donations, or the very *content* of on-campus speech. The university reserved the right to approve content (to assure it was "appropriate"), and it banned the recruiting of members to partisan causes. It may be no accident that this change in the status of the space in front of Sproul Hall occurred just as political activism was heating up in reaction to the conservative but benign hegemony of the Eisenhower administration and the continuing and far less benign actions of the House Un-American Activities Committee and entrenched anti-civil rights racists in both the South and the North.

Three issues had emerged by the late 1950s that made the administration wary of allowing political activity on the land that it controlled. Within the university, there was increasing agitation to abolish compulsory ROTC (Reserve Officer Training Corps) for male students. Within California, both state and national House Un-American Activities Committees were becoming more aggressive again after a slight lessening of activity in the mid-1950s.[9] And nationally students were becoming active in the civil rights movements in the South and in other liberal and leftist causes, and they were beginning to bring that activism back to their campuses in the form of the demand that the same rights being agitated for in the South be extended to students at the university. The increasing assertiveness of students on these issues, coupled with a University of California administration (and Board of Regents) that was increasingly defining the role of the university as an institution in service of the economy and the society, suggested to university officials that clear guidelines on "appropriate political behavior" of students needed to be established.

Since 1938, student political activity had been guided by "Rule 17," which required presidential approval of off-campus speakers and for the use of university property by nonrecognized groups.[10] Additionally, Rule 17 forbade the collection of funds on campus by any student or nonstudent group. In October 1957, a new "liberalized" interpretation of Rule 17 was offered by the university administration in response to a year's agitation by various student groups. The new interpretation, which became a center of controversy in 1964, allowed off-campus groups composed entirely of students to use campus facilities provided that the dean of students approved the use at least a week in advance. All promotional material also had to be cleared through the dean. Any necessary services, such as police protection (which the university required), were to be paid for by the sponsoring group. Off-campus speakers no longer had to be approved by the president, but they did have to be approved by the dean, a faculty or senior staff advisor, and occasionally an appropriate departmental chair. Finally, off-campus groups were not allowed to solicit for either funds or membership (Heirich and Kaplan 1965, 19). It was obvious that the administration felt that it was its right and duty to continue to monitor closely the political activities of the students in the UC system, even if such monitoring was now removed from the president's office. At the same time, as the campus spread south, the traditional "free speech" area was abolished, although political activity was allowed on a strip of sidewalk opposite Telegraph Avenue on Bancroft Way (Figure 3.4). With this arrangement, speakers were presumably off university property, but audiences at rallies and speeches often spilled into Sproul Plaza.

The 1957 liberalization of restrictions on political activity were soon tightened back up when, in the fall of 1959, UC President Clark Kerr released what came to be known as the Kerr Directives. The Kerr Directives forbade student governments from speaking on "off-campus" issues, made the governments and student organizations directly responsible to the chancellor's office on each campus, provided that any amendments to government or organization constitutions be approved by campus officials, and required that all student organizations have a tenured faculty advisor (Heirich and Kaplan 1965). In October 1960, the UC administration arranged to have the editor of the student newspaper, *The Daily Californian*, removed for supporting student government candidates who were opposed to the policies of the administration on issues of free speech and the ROTC. The following April, the chancellor of the Berkeley campus issued a new set of rules that prohibited persons "unconnected to the university" to post, distribute, or exhibit literature on campus. Throughout the next 2 years, the university administration at both the campus and system level was engaged in constant clarifications and reclarifications of what appropriate on-campus political activity was. Most consistently, throughout these constant revisions, the administration reserved for itself the right to control both the content and the form of political activity on campus.

The development of student political consciousness on campus, and the continued attempt by the UC administration to maintain and solidify its control over political activity on campus, occurred concurrently with a series of social changes in the South Campus area, changes

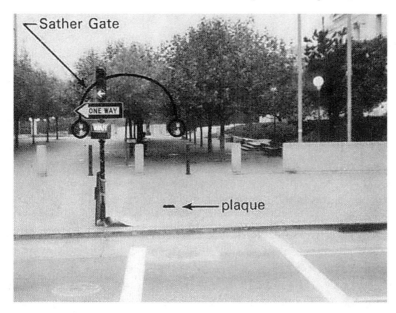

FIGURE 3.4. A photograph indicating one of the plaques on the sidewalk along Bancroft Avenue at Telegraph Avenue. The area in front of the plaque is city property. Behind the plaque, stretching to Sather Gate, is the portion of Telegraph Avenue ceded by the city to the university when the student union was built. Before the Free Speech Movement, speakers would often stand on the city portion of the sidewalk and speak to crowds on university property. This is one of the practices the university sought to halt in the fall of 1964. Photograph by Lyn and John Lofland, originally published in Heirich (1971); used by permission.

that the administration saw as at least as threatening as those posed by students demanding a political voice. At the university's request, the city of Berkeley had dutifully conducted a study that declared the South Campus area to be "blighted"—a blight made all the more menacing by the realization that "Telegraph Avenue [had come] to rival San Francisco's North Beach as the vital center of the Beat Generation . . . " (Scheer 1969, 43). The idea that South Campus was blighted was reinforced by the growing "counterculture" centered on Telegraph, a counterculture that seemed to be as pernicious as it was attractive to students and other youths. Robert Scheer rather caustically remarked after the People's Park riots of 1969 (discussed later in this chapter) that the South Campus area, by the early 1960s, had come to be understood by the authorities of California as a watering hole gone bad. . . . Perfectly decent young men and women attending what was supposedly the star attraction of the whole state university network were turning out to be politically and socially deformed, causing trouble for parents and politicians alike. And it all seemed to have something to do with a place called Telegraph Avenue where "they" practiced fornication, smoked marijuana, wrote leaflets, mobilized protests, and read sinister revolutionary tracts. (Scheer 1969, 43–44)

A more "sober-minded" analyst, and an opponent of Scheer's, suggested essentially the same thing. Quoting Max Weber, Seymour Lipset argued that students, precisely because they were young, lacked an ethic of responsibility: they were not accountable for the consequences of their actions. As Lipset wrote in the wake of the Free Speech uprisings: "University students, though well educated, have generally not established a sense of close involvement with adult institutions; experience has not hardened them to imperfection. Their libidos are unanchored . . . " (Lipset 1965, 9).[11] And Heirich (1971) later argued that the explosive combination of environmental change (on campus) and environmental disorder (in the South Campus area) with youthful segregation and premature autonomy were responsible for what he called the "unreasonable" nature of protest in Berkeley in 1964. The transformation of South Campus into a haven for "beats" and student organizing suggested to the university that *in loco parentis* was breaking down, and the university was at a loss to explain its demise. By the mid-fall of 1964 it actually didn't much matter if the university could explain what was happening or not, for by then it was fighting a rear-guard action against the wild youths with their unanchored libidos—or more accurately against a committed group of politically savvy and well-organized students who were quickly gaining support from the larger masses of their heretofore less politically active colleagues. The crisis, however, was of the university's own making.

The Geography of Free Speech 2: The Free Speech Movement

On September 16, 1964,[12] all student organizations received a letter from the dean of students, Katherine Towle, informing them that the 26-foot strip of sidewalk along Bancroft Way, which had become the de facto free speech area when Sather Gate was engulfed by the campus, would no longer be available for proselytizing and fund-raising (see

Figure 3.4). The strip of land was legally university property and as such was subject to the same regulations and restrictions as other parts of campus. The university justified its actions by pointing out that it had lifted a ban on scheduled outside speakers and had established a "Hyde Park" area as an open forum for students and staff in the plaza below the Student Union (Figure 3.5).

The problem with the university's new "Hyde Park" area was that it was, quite literally, out of the way. For exactly that reason it was unacceptable to students and their supporters in the community and among staff and faculty, even though it was seen as a convenient solution by the administration. Responding to this new and geographic restriction on public speech, students, working through organizations as diverse as the leftist CORE (Congress of Racial Equality), SNCC (Student Nonviolent Coordinating Committee), and SDS and the right-wing Young Republicans and Students for Goldwater, protested and engaged in a program of open defiance of the ban.

FIGURE 3.5. The plaza below the student center that the administration designated in the midst of the Free Speech Movement to be a "Hyde Park area." Even with a pub featuring outdoor seating at the edge of the plaza and the Zellerbach Auditorium concert hall, the plaza remains a place where relatively few people gather or linger. Photograph by author.

Early entreaties to the Berkeley administration by the united student groups asked for the reinstatement of students' and others' right to set up tables and distribute political literature on the sidewalk at Bancroft and Telegraph Avenues. Student leaders also announced plans to contact lawyers who would consider taking legal action against the university. Dean Towle hinted that political leafleting and tabling might be allowed in the existing "Hyde Park" area, but students once again reiterated the unacceptability of the lower Sproul Plaza as a *political* space. On September 18 a coalition of 18 student organizations presented to Dean Towle, in the form of a petition, what amounted to a set of "time, place, and manner" rules to govern the Bancroft–Telegraph sidewalk. This petition was rejected. In response, on September 20, the students voted to engage in a course of civil disobedience if the university remained firm in its ban on political activity after a meeting with the dean the following morning.

On September 21, Dean Towle acceded to many of the students' demands—but not all of them. She announced that tables and leafleting would be allowed on the Bancroft–Telegraph sidewalk but that only "informative" (and not "advocative") literature could be distributed; that fund-raising would not be permitted; and that "recruiting" people to organizations would not be tolerated. As Towle put it: "It is not permissible, in materials distributed on University property, to urge a specific vote, call for direct social or political action, or to seek to recruit individuals for such action."[13] Simultaneously Dean Towle announced that a "second" Hyde Park area would be established—on an experimental basis—on the steps of Sproul Hall. Here only students and university staff could speak: "Since the university reserves such areas of the campus for student and staff use, those who speak should be prepared to identify themselves as students or staff of the university."

The students rejected the concessions and announced plans to engage in civil disobedience. As one student organizer, Jackie Goldberg, put it:

> [T]he University has not gone far enough in allowing us to promote the kind of society we're interested in.

> We're allowed to say why we think something is good or bad, but we're not allowed to distribute information as to what to do about it. Inaction is the rule, rather than the exception, in our society and on this campus. And, education is and should be more than academics.

We don't want to be armchair intellectuals. For a hundred years, people have talked and talked and done nothing. We want to help the students decide where they fit into the political spectrum and what they can do about their beliefs. We want to help build a better society.

Dean Towle argued that the "nonadvocacy" position was part of university-wide policy and as such was something the Berkeley administration was powerless to change. About 75 students, unswayed by this logic, held an all-night vigil on the steps of Sproul Hall.

Other students, working through the Senate of the Associated Students of the University of California (ASUC), petitioned the Board of Regents the next day "to allow free political and social action to be effected by students at the Bancroft entrance to the University of California, up to the posts accepted as the traditional entrance." Open defiance of the nonadvocacy provisions announced by Dean Towle began. On September 27, in part as a response to an unforgiving statement by UC President Clark Kerr, students announced that the following day, during a University Meeting, they would establish tables on the sidewalk at Sather Gate and hold a rally at Wheeler Hall without properly notifying the administration.

At the September 28 University Meeting, Berkeley Chancellor Edward Strong announced a number of concessions to the Free Speech protesters. Among others, these concessions included allowing limited forms of advocacy (e.g., promoting a "yes" or "no" vote on initiatives, and distributing campaign bumper stickers and buttons). Students interpreted this reversal from the policy announced by Dean Towle only a few days earlier as a direct result of their picketing and rallying. The next day, a number of groups set up tables both at Sather Gate and at Bancroft–Telegraph. Only a few of these groups had secured the proper permits from the dean of students. Under the new policy announced by Chancellor Strong the day before, only groups that "promised not to solicit money or members, or initiate or advocate any off-campus activity other than voting" would be issued permits, and most groups simply refused to make this promise.

The following day, September 30, 1964, the situation exploded. In the early afternoon, five students staffing tables were requested to appear before the dean of men at 3 P.M. for violating university regulations: none had permits and some were collecting money for off-campus political activities. More than 600 students quickly signed a statement saying that they had been equally responsible for staffing tables and that they too should be required to meet with the dean of men. At 3 P.M., some 300–500 students appeared outside the dean's office in Sproul Hall, with some, including the soon-to-be-famous Mario Savio, Arthur Goldberg, and Sandor Fuchs, taking up a position on an exterior balcony and exhorting passing students to join the demonstration.

In response to the demand that all those who had signed the statement claiming to have violated university policy be treated equally, the dean of men responded that the administration would cite only those "observed" breaking university policy, but he agreed to meet with the five who had been cited plus Savio, Goldberg, and Fuchs at 4 P.M. All eight refused to appear, and students decided to continue occupying Sproul Hall through the night. Around midnight Chancellor Strong issued a statement first asserting that UC students were more free than any others to engage in political action and then indefinitely suspending all eight students. In the early hours of the morning, after christening themselves as the Free Speech Movement, the occupiers of Sproul Hall ended the sit-in. Student organizers, with Savio as their spokesman, announced a rally for noon that day, October 1, on the steps of Sproul Hall.

As organizers were posting flyers announcing the rally, two tables were set up on Sproul Plaza at the bottom of the Spoul Hall steps. One of those tables was staffed by Jack Weinberg, a former student. When two deans asked him to provide identification, Weinberg refused to do so. He also refused to leave the table, whereupon a police lieutenant accompanying the deans arrested him. Students in the area protested, chanting "release him, release him," and perhaps two hundred lay down on the pavement all around the police car he was being taken to so that it could not leave Sproul Plaza. After Weinberg was placed in the car, Mario Savio climbed on its roof (after first carefully removing his shoes) and implored students and others in the area to join the protest (Figure 3.6). Students maintained their vigil around the police car—with Weinberg inside it the whole time—for 32 hours. A rotating group of student leaders climbed to the top of the police car to make demands upon the university, while a phalanx of protesters reoccupied Sproul Hall. When campus and city police tried to close Sproul Hall at about 6:15 P.M. on the first day of the standoff, about 2,000 protesters rushed the doors, knocking at least two police officers out of their way, and occupied the hall in an uneasy standoff with police. Some hours later, at the request

FIGURE 3.6. Mario Savio addressing the crowd from the roof of the police car that held Jack Weinberg, October 1, 1964. Photograph by Ron Enfield.

of students gathered in the plaza, those in the hall returned outside and relinquished the building to the police and administration, where protest leaders, working in a closed session, worked out a plan for ongoing civil disobedience at least through October 3, UC Berkeley's "Family Day."

Despite counterprotests by those opposed to the student activists (and the kindling of a near-riot as contending groups jostled with one another), and despite growing cracks in the cross-ideological coalition that had formed originally to protest restrictions on speech and political activity, Free Speech activists maintained their vigil at the police car. Governor Pat Brown announced his support of the university and campus administrations, and Chancellor Strong announced that the protests, in fact, were not about free speech: "Freedom of speech by students on campus is not the issue. The issue is one presented by deliberate violations of University rules and regulations by some students in an attempt to bring about a change of the university policy prohibiting use of University facilities by political, social and action groups." Consequently, Strong and UC President Kerr determined during the morning of October 2 to attempt to regain control of the plaza. With the support of the governor's office, Strong and Kerr agreed that at 6 P.M. that evening the protest would be declared an unlawful assemblage, and if protesters did not voluntarily disperse, police would force them out. By 4:45, some 500 police officers from a range of Bay Area and state authorities marched to the campus and took up positions near Sproul Hall. The protest crowd grew in response—to perhaps as large as 7,000. A confrontation seemed likely.

Chancellor Strong and President Kerr agreed to meet with activists at 5 P.M. in advance of the 6 P.M. announcement. At about 5:30, the crowd was informed that the president had delayed any police action while a meeting with protest leaders, clergy, and faculty members was in session. At 7:15 the meeting disbanded, and at 7:30 Savio climbed atop the stranded police car and read the agreement:

1. The student demonstrators shall desist from all forms of their illegal protest against University regulations.
2. A committee representing students (including leaders of the demonstration), faculty, and administration will immediately be set up to conduct discussions and hearings into all aspects of political behavior on campus and its control, and to make recommendations to the administration.
3. The arrested man will be booked, released on his own recognizance, and the university (complainant) will not press charges.
4. The duration of the suspension of the suspended students will be submitted within one week to the Student Conduct Committee of the Academic Senate.
5. Activity may be continued by student organizations in accordance with existing University regulations.
6. The President of the university has already declared his willingness to support deeding certain University property at the end of Telegraph Avenue to the city of Berkeley or to the ASUC.

Savio urged the protesters to end their occupation of the plaza and to go home "with dignity." The protesters assented and the demonstration broke up.

Meanwhile, President Kerr held a press conference confirming the details of the agreement and announcing that Chancellor Strong would set up the ad hoc committee mentioned. He also noted, however, that the UC administration would not be bound by the recommendations of the ad hoc committee: they were to be recommendations only. Finally, he stated that while the university would not press charges, he could not speak for the district attorney, who might (and in fact did).

Over the next several days both the administration and the protesters skirmished—verbally, at least—over the meaning of the October 2 agreement, with students holding a large, and illegal, rally at Sproul Plaza on October 5. The activists agreed at the rally to suspend political activity in the contested spaces until after the ad hoc committee met and formulated its policy recommendations. When Chancellor Strong soon afterward announced the members of the ad hoc committee, elected FSM leaders immediately protested, saying they had not been consulted in the manner that they felt the October 2 agreement required them to be. When the committee met for the first time on October 7, ten FSM leaders appeared before it, declared it to have been illegally constituted, asked it to disband, and walked out. For its part, the committee announced itself to be a "study" rather than a "policy" committee and, after much discussion, determined to hold hearings on campus political activity beginning in 1 week.

Simultaneously, Clark Kerr went on a public relations offensive, declaring that though students were more activist than ever, the Berkeley protest was "one episode—a single campus, a small minority of students, a short period of time"—that is, an aberration. He more than once went out of his way to note that some of the demonstrators had "communist sympathies." He also reiterated that the administration was acting within the spirit of both the October 2 declaration and the negotiations that led to it, a position that received some support from faculty members who had brokered much of the agreement. Chancellor Strong also went on record, declaring the protests to be the result of "hard-core protesters" who wanted to "open up the university," and that his administration was determined to make sure that "the university will not be used as a bastion for the planning and implementation of political and social action." Despite and because of these pronouncements, Kerr, Strong, and both the UC and Berkeley administrations found themselves buffeted by continual and often competing representations, petitions, complaints, and threatened protests—from various student groups such as fraternities, sororities, and even 29 Oski Dolls (UC Berkeley cheerleaders), as well as from ad hoc faculty groups, state politicians, and newspaper editorialists.

When the Study Committee on Political Activity (as the ad hoc committee was renamed) held its first public hearing on October 13, all but one of the 300 speakers rose to declare the committee to be illegally constituted. Partially in response, a new agreement was forged between the FSM steering committee, the administration and other interested parties. Announced on October 15, the agreement reestablished the Study Committee on Political Activity along new lines: it was enlarged from 12 to 18 members; exact means by which members were to be appointed were specified; provisions were made for twice-weekly hearings to last for 3 weeks; two attorneys and five other "silent observers" were invited to join the hearings; and it was determined that all decisions were to be made by consensus.

As both the Committee on Political Activities hearings and separate hearings into the suspension of the eight originally cited students got under way, FSM leaders debated the efficacy of continuing a ban on political tables on campus and decided, despite the impending national election, to maintain the ban while the committees did their work. After the election, on November 9, however, feeling that the process was moving too slowly, and in response to administration arguments that it needed to retain the right to discipline students or organizations that advocated acts that "directly result[ed]" in "unlawful acts" off campus, FSM-affiliated groups returned to "tabling" on Sproul Plaza and at Bancroft–Telegraph. The FSM argued that the determination of whether activities were illegal was up to the courts to decide, not the university administration, and that it needed to "exercise its constitutional rights." The time to test the administration's position on the use of the campus for speech had come.

The following day, some 70 students received letters citing them for violating (still in effect) university policies. Once again, hundreds of students (many of them graduate students, who by this time had also begun to explore the possibility of unionizing) signed petitions claiming equal responsibility for breaking university regulations. Despite the citations, the university took no action against the people staffing the tables and allowed them to continue their advocacy work. On November 20, as the University Regents were meeting in University Hall, some three thousand students rallied at Sproul Hall before working their way to the Regents' meeting. At the meeting, FSM and other student representatives were barred from speaking. The Regents eventually voted, on President Kerr's recommendation, to adopt a modified version of regulations developed by the Committee on Political Action to allow fund-raising and recruitment, but banning "illegal advocacy." At the same time, the Regents more or less rejected a faculty committee recommendation that the originally cited students only be "censured" and instead reinstated them without clearing their records. In response, graduate students called for a sit-in, but Savio argued, successfully, for a cooling-off period over the weekend, followed by a rally on Monday, November 23.

On that day, several hundred students reoccupied Sproul Hall, but only after a fierce debate within the FSM (that by many accounts "split" the movement). After the Thanksgiving weekend, and as many FSM activists reestablished tables on the plaza and Bancroft–Telegraph, graduate students voted to strike, beginning on December 4. In the meantime, three FSM leaders, Savio, Arthur Goldberg, and Jackie Goldberg, received letters saying that new disciplinary charges stemming from the October 1–2 protests were being lodged against them by the administration. On December 2, 800 students once again occupied Sproul Hall to protest the administration's "arbitrarily singling out students for punishment" and what they saw as a continuing refusal to negotiate in good faith.

Governor Brown responded on December 3 by sending more than 600 police officers to Sproul Hall to arrest the demonstrators. Arrests lasted more than 12 hours. Sympathetic students and faculty staged a spontaneous strike. Taking their strongest stand yet, some 900 faculty members met that night and called for complete amnesty for the FSM protesters and for complete and unconditional political freedom for students—including the right to engage in advocacy. Departmental chairs working behind the scenes tried to meet with the administration to negotiate a settlement but were rebuffed. The next day, as the strike continued—quite effectively—and as the administration maintained an eerie silence, refusing to talk with faculty or departmental chairs, the chairs of all campus departments constituted themselves as the Council of Chairmen in hopes of reestablishing at least some authority on campus (since the general sense was that both the campus and system administrations had pretty much abdicated).

Frenetic rounds of negotiation followed, as nearly all normal campus activity ground to a halt. Over the weekend of December 5 and 6, the Council of Chairmen met in long sessions to work out a plan to end the protests, and after a meeting between the head of the Council of Chairmen and President Kerr, and later between Kerr and the Regents in a South San Francisco motel, an agreement was reached. Classes were cancelled campus-wide on Monday morning, December 7, so that departmental meetings could be held to discuss the agreement. In brief, the agreement represented a significant victory for the FSM: complete amnesty was granted to protesters for actions through December 7, and no position by the administration was to be taken on the question of advocacy. Following the departmental meetings, a huge convocation was held in the Greek Theater to announce the terms of the agreement (Figure 3.7). At the conclusion, Mario Savio attempted to speak to the assembled students and staff but was pulled from the stage by police officers. When he was finally allowed to speak, he announced a rally for noon at Sproul Hall.

At the noon rally, department chairs and FSM leaders announced the end of the strike while the Academic Senate considered the proposal for complete political freedom and the right to advocacy. Rumors quickly spread that in closed-door meetings President Kerr had agreed to the opening up of the campus for political activity. The following day, the Academic Senate voted 824–115 to accept a resolution allowing political speech and advocacy on campus and lifting restrictions on students' off-campus political activities as well as political activities by nonstudents on campus. The Senate resolution noted, however, that the Senate, as the lawmaking body on campus, needed to regulate the "time, place, and manner" of speech activities on campus, in essence returning the university to the *status quo ante* of 1938 before "Rule 17" had been implemented—and aligning the campus with other publicly owned spaces in the city. The Free Speech Movement—at least in its directly activist form—was over. A significant victory in favor of students' rights of assembly and speech—and of control over their campus—was won.

The Geography of Free Speech 3: The Where of Protest

A fight over location, coupled with a fight over "appropriate" forms of speech and political action, proved to be explosive not just for the campus but for Berkeley and beyond. When Berkeley students and activists

FIGURE 3.7. Clark Kerr addresses the meeting at the Greek Theater on December 7, 1964 (top). The meeting was called to announce the terms of the agreement ending the Free Speech protests. When Mario Savio sought to address the crowd at the end of the meeting, he was pulled from the stage by policemen (bottom). Photograph originally published in Heirich (1971).

won the right to set up tables and promote political action on the Telegraph–Bancroft sidewalk and in Sproul Plaza, they in essence won the *right* to a particular space—the campus. From that space, many sought to organize a new kind of society, a new kind of city. But make no mistake, control of *a* public space was crucial, since, after all, it was only through control over that space that political action could expand. The

Berkeley campus became what Bruce D'Arcus (2001) calls a "protest platform"—something akin to a "liberated zone" from within which political action could be organized.

This was an issue clearly grasped by both FSM activists and the UC administration. As Clark Kerr had already said, it seemed as if many of the students hoped to turn the campus "on the Latin American or Japanese" model into a staging ground for radical societal transformation. As we will shortly see, they were, in fact, to some extent successful. But first it is important to emphasize just how much this was a *locational* conflict. The Free Speech Movement began as a response to the university's attempts to control or direct the speech activities of its students (and others who used the campus). The argument was that the *institution* of the university controlled and had full rights over the *space* of the university, that the campus was simply not a public forum in the traditional sense. In terms of the public forum doctrine that was even then emerging at the level of the Supreme Court, the university argued, at least implicitly, that the campus was at best a "dedicated" public forum and thus it had a right to more closely regulate the types of speech activities engaged in, their specific locations, the times they could occur, and so forth. The campus simply was not a city street or park and was not to be treated like one. By contrast, activists argued, again implicitly, that the campus was in fact a traditional public forum (or should have been one) and that the university had no right to regulate speech beyond the regulations already provided for in law. For these activists, there was no clear distinction between a city-owned sidewalk and a university-owned one, except insofar as the university-owned one was a better location for their activities. Activists worked to assert their right to the *particular* space of the campus (as opposed to simply moving their activities onto city-owned property).

At a finer spatial scale, the Free Speech Movement was even more a conflict over location. The university's establishment, and the activist's rejection, of a "Hyde Park" area in the lower Sproul Plaza indicates just how much the Free Speech Movement was concerned with the question of *where* protest or other political activity should be located. For the students, to "go again to Hyde Park" meant something entirely different than it did for the administration. For the administration, it meant that certain "Hyde Parks" convenient to *it* could be established. For activists it meant retaking the *prime* protest and political locations of the university and city. It meant reclaiming the sidewalk at Telegraph and

Bancroft. It meant establishing a right to the plaza at the foot of Sproul Hall (and just outside Sather Gate)—the traditional heart of the campus. Indeed, it meant reclaiming the steps of Sproul Hall themselves. It meant *taking* a space and *making* it public (a point to which I will return in greater detail in the next chapter).

To this day, Sproul Plaza remains a prime political space on the Berkeley campus. Nearly every lunchtime, activists set up on the Sproul steps and address the passing crowds. Along the walk to the Sather Gate, numerous organizations—both "on" and "off" campus ones—set up tables and distribute literature. On important occasions, marches and rallies are organized or held in Sproul Plaza. It remains a vibrant space for politics.

FROM FREE SPEECH TO COUNTERCULTURE: URBAN RENEWAL AND THE BATTLE FOR PEOPLE'S PARK

Following the victories of the Free Speech Movement, the transformation of the neighborhoods around the Berkeley campus intensified. The movement proved to be a great reinforcer of the bourgeoning counterculture of the South Campus area (Scheer 1969). And, just as President Kerr had feared, the campus itself became something of a "free zone" for political activists. The Vietnam Day Committee, among others, was accused by relatively conservative faculty and others of using the university as a "staging ground" for subversive forays into the larger community. The otherwise liberal philosopher and sociologist Lewis Feuer, in particular, was deeply outraged by the Free Speech Movement and its effects on the Berkeley campus. Feuer blamed the faculty for refusing to "properly" limit the rights of students. This refusal had allowed the campus to "safeguard the advocacy and planning of immediate acts of violence, illegal demonstrations, interferences with troop trains, and obscene speech and action" (Feuer 1966, 78). The value—to activists—of a "liberated" staging ground—a public space—could not be clearer. The value—to liberalism—of *order* in public spaces also could not be more clear.

Perhaps that is the reason that the university, too, wanted to use the campus as a staging ground for an assault on the urban fabric of the South Campus area, even as city plans for urban renewal were coming under increasing fire by merchants and residents of the area. In the spring of 1966, public hearings, required by federal law, were held on the city's Long-Range Development Plan, which called for extensive urban renewal and redevelopment in South Campus and along Telegraph Avenue. Opposition to the plan was strong enough that the Berkeley *Gazette*, a supporter of city-wide redevelopment, had to admit that the residents of the South Campus area "do not now, and have not in the past, liked the plan." But support from outside the district was strong from those who were "aghast" (as the *Gazette* put it) at the "beatnik" development that appeared to be arising in the absence of a strong city program of redevelopment. During a series of delays in implementing the plan, an uneasy alliance of students, local merchants worried about increasing rents, and older people "living in lifetime homes" in the South Campus area organized effectively enough to defeat the redevelopment plans (Scheer 1969, 44).

Despite the demise of the Long-Range Redevelopment Plan and the end of comprehensive urban redevelopment in Berkeley, the university maintained an aggressive desire to expand into the South Campus area. In particular it eyed a series of lots, mostly occupied by relatively run-down older houses, for dormitories and other "nonacademic uses" over which the university would nonetheless have control. As part of this expansion plan, the university in 1967 authorized its new chancellor, Roger Heyns, and Vice Chancellor Earl Cheit to purchase lot 1875-2 between Dwight and Haste Streets (see Figure 4.2, page 121). Original university plans had called for the construction of high-rise dormitories on this site, but because vacancy rates were at an all-time high in the city, dormitories were not really feasible. The university thus announced that the purchase of the site, and the clearing away of the houses on it, was designed to address a "desperate need" for a new soccer field in the area. When pressed by a reporter, the chairman of the campus Building and Development Committee admitted that one of the effects of the purchase would be to transform the South Campus neighborhood—to assist in eliminating the "counter culture" that had begun to grow up around, and define, the university: "I presume it is true. You are killing two birds with one stone. But we are aiming at only one of them; the other is free. We are seeking more facilities and if you engage in urban renewal, that's an added benefit" (Scheer 1969, 44).[14] University Regent Fred Dutton remarked, after the decision to buy the property had been made, that Heyns and Cheit had presented their plan for the lot to the Board of Regents as "an act against the hippie culture" (Scheer 1969, 44).

In 1967 the campus administration bought the land—through the imposition of eminent domain—despite the fact that no funds were appropriated to improve it once purchased. The resolution that justified the purchase left no room for disagreement over the reason the purchase had been approved: "The Regents have approved the use of $1.3 million in U.C. funds to purchase three acres south of the Berkeley campus. The area has been a scene of hippie concentration and rising crime." In June 1967, still without money for improvement, the houses on the three acres that comprised lot 1875-2 were demolished (Scheer 1969, 46).

All through 1968 and into 1969, lot 1875-2 remained unimproved—a muddy patch of ground that had become a free parking lot and, to many in South Campus, a symbol of the contempt in which the university held them. Indeed, it sat like a hole right in the heart of what was fast becoming the center of political and cultural transformation in Berkeley. While Sproul Plaza still remained a vital center for organizing, Telegraph Avenue had increased in importance as a site for experimentation, political meetings, neighborhood solidarity, and anti-war activism. During the summer of 1968, Telegraph Avenue was the scene of a series of pitched battles between riot police and antiwar demonstrators. And on campus, during the winter term of 1969, a wide coalition of students called for a strike (not the first since the FSM, either) to win their demands for a range of ethnic studies programs. The strike met with a good deal of success. Fearing an escalation of the occasional violence that had marked recent demonstrations, Chancellor Heyns turned over command of the campus police to (the notoriously tough) Alameda County Sheriff Frank Madigan, and requested that the new governor, Ronald Reagan, declare a state of emergency. Reagan readily agreed and, as the so-called Third World Strike quickly withered in the face of severe police brutality (Lyford 1982, 38), police forces were gradually withdrawn. The state of emergency, however, still remained technically in effect.

In this context, the following announcement appeared in the Berkeley *Barb* in April 17, 1969:

A park will be built this Sunday between Dwight and Haste.
The Land is owned by the university which tore down a lot of beautiful houses in order to build a swamp.

The land is now used as a free parking space. In a year the university will
build a cement-type expansive parking lot which will compete with
the other lots for the allegiance of the Berkeley Buicks.

On Sunday we will stop this shit. Bring shovels, hoses, chains, grass,
paints, flowers, trees, bull dozers, top soil, colorful smiles, and lots
of weed. . . .

We want the park to be a cultural, political freak-out rap center for the
Western World. . . .

This summer we will not be fucked over by the pigs "move on" fascism,
we will police our own park and not allow its occupation by an imperial power. . . . (reprinted in Lyford 1982, 40–41)

Activists, in other words, were planning to take (or perhaps take back)
another space in the name of creating an open community-controlled
political space. They were planning to make a People's Park (Figure
3.8).

Although the reasons for being involved in the park were as varied
as the people who turned out on that first Sunday morning, April 20,
1969, there still was the understanding that the construction of the park
was a symbolic act that struck at both the designs of the university as a
capitalist enterprise (in the terms long before articulated by Clark Kerr)
and at capitalist society itself.[15] As one of the park supporters recalled a
few years later: "The builders of the park were not a gang of ideological
do-gooders. . . . Although economic and environmental issues were
raised by park developers and supporters, fundamental to the struggle
was the right of ownership, and the nature of private property rights"
(quoted in Lyford, 1982, 41). Robert Scheer (1969, 46) suggested that
most of the builders were of the "nonsectarian breed that managed to
get through Berkeley's ideological warfare with a sense of humor and
spontaneity in tactics." Soon People's Park became an "event." On
weekends as many as 3,000 people worked at planting flowers and
building playgrounds (Figure 3.9). The development of the park had
broad-based support on campus and within the community. A letter to
the *Daily Californian* protesting the university's decision to reclaim the
land the park was built on was signed by 84 students leaders, including
not only activists but also fraternity presidents, the head of the pom-
pom girls, and the leader (again) of the Oskie Dolls.

Despite such support for People's Park, Chancellor Heyns decided
that the university could not simply ignore such a strong challenge to
its authority—and its ownership of lot 1875-2. On May 14, 1969, before

FIGURE 3.8. The muddy parking lot (lot 1875-2) that eventually became People's Park. Photograph by Mark Harris.

leaving town on business, Chancellor Heyns privately ordered the park to be cleared of any "residents" and a fence to be built around the perimeter. Heyns arranged for the Alameda County Sheriff to provide protection for the work crews that would remove people from the land and build the fence. Since Vice Chancellor Cheit was also out of town, a second vice chancellor was left in charge of the operations, although he later claimed that he had not been told the fence was to be constructed. He claimed that he "was told not to expect any problems" (quoted in Scheer 1969, 52).

At first it looked as though there just might not be any problems. The fence was constructed at 5:30 A.M. without any disturbance. The rationale for this action was presented by Chancellor Heyns a day later:

We have been presented with a park we hadn't even planned or asked for. . . . So what happens next? First we will have to put up a fence to reestablish the conveniently forgotten fact that the field is indeed the university's, and to exclude unauthorized persons from the site. That's a hard way to make a point, but that's the way it has to be. (quoted in Lyford 1982, 43)[16]

The first protest against the fence was called for noon on May 15—to be held in the long-since-liberated Sproul Plaza. About 6,000 protesters massed and, urged on by just-elected student body president Dan Siegal to "reclaim the park," began to march down Telegraph Avenue.

There they met the arrayed forces of the Berkeley city police and the Alameda County sheriffs, who attempted to disperse the crowd. Under a giant billboard proclaiming "Showtime" (Figure 3.10), the rioting that ensued was vicious and bloody. At least 128 protesters were injured, one was blinded after being shot in the eyes with buckshot, and one other—James Rector—was fatally wounded as he watched the riot from a roof above Telegraph Avenue. No police officers were seriously injured. Alameda Sheriff Frank Madigan, whose officers were responsible for most of the injuries and Rector's death, defended the use of force by claiming that the crisis had been instigated by "anarchists and revo-

FIGURE 3.9. Building the Park. Hundreds of people turned out on successive weekends to construct People's Park. People brought tools, donated materials, or simply provided labor, as their means permitted. Photograph by Jean Raisler.

lutionaries" intent on taking "this form of government down, starting with the educational system and then with law enforcement" (quoted in Lyford 1982, 53). That night, referring to the continuing state of emergency, Governor Reagan remobilized the National Guard and banned public assemblies.

Despite the ban, protesters—students, city residents, sympathetic faculty members—gathered daily on and off campus to protest both the fencing of the park and the ongoing use of force by the police and National Guard, including the famous tear gas "bombing" of the campus by National Guard helicopters (Figure 3.11). Following a show of overwhelming support for the park in a campus-wide referendum, Chancellor Heyns announced on May 29 that he supported leasing the land to the city of Berkeley. The next day some 30,000 people march peacefully past the park as violence subsided. The fence, however, did not come down. Indeed, it remained under 24-hour guard.

After the Riots

On June 20, 1969, Governor Reagan pushed a proposal through the Board of Regents that called for the construction of student housing on the site of People's Park, a return to the original Long-Range Plan of

FIGURE 3.10. Showtime. The People's Park riots. May 1969. Photograph by Ed Krishner.

1952. In doing so, Reagan engineered the rejection of a compromise plan, supported by the chancellor, that would have leased the land to the city of Berkeley for 7 years with provisions for the maintenance of a user-constructed park on at least a portion of the parcel (Scheer 1969, 53). Whereas the builders of the park and its defenders saw the park as an unalienated space for social, cultural, and political action, Reagan, echoing Matthew Arnold from so many years ago, saw things rather differently: the disturbances in Berkeley (and, in sympathy, throughout the university system) were not "simply the acts of youngsters sowing their wild oats or legitimately questioning our society and its values" (quoted

FIGURE 3.11. The famous teargassing of the Berkeley campus during the People's Park riots. The student center is in the foreground. Most of the teargas drifted north of Sather Gate into the main part of campus and beyond into the wealthy residential neighborhoods to the north. Photograph by Andrew R. Scott.

in *Los Angeles Times*, 1969). While more than a panty raid, the protests—including the taking of the land in the first place—were less than legitimate. Presumably, building another dorm would help reassert control over the students and other rioters.

For park builders and protesters, of course, this had never been a "panty raid." It was, in fact, a much more fundamental fight. Lot 1875-2 became a symbol of the arrogance and the power of the university (which itself stood as a symbol of "the system," or "the establishment"). Throughout the 1950s and 1960s, the university had claimed for itself, and attempted to enforce, the right to determine the nature and form of political discourse. Against this, students and others were struggling to find new and (in their eyes) appropriate forms of expression. Doing so required the taking, occupation, and radical transformation of space: it necessarily led to conflicts over location (where political speech could occur; where dorms should be built) which were at the same time struggles over rights (who had the right to speak; who had the right to determine the fortunes of whole neighborhoods). Making People's Park and subsequently defending it, like the Free Speech Movement that preceded it, were experiments, certainly imperfect, in the radical democratization of decision making, and of the adjudication of conflicting rights—including, quite apparently, the right to the campus and to the city—in Berkeley.

In the end, Governor Reagan denounced the protesters as "street gangs," asserting that they were a "well-prepared and well-armed mass of people who had stockpiled all kinds of weapons and missiles" (quoted in *Los Angeles Times*, 1969). And despite a second riot in 1971, the fence remained around the park until May 1972, when protesters ripped it down in reaction to President Richard Nixon's announcement that the United States was planning to mine harbors in North Vietnam.

But if the disposition of lot 1875-2 remained unsettled, it nonetheless served as a rallying point for the transformation of politics in Berkeley. A coalition of radical groups began to organize around a series of electoral issues, and People's Park became a symbolic center: what had been a battle over a specific space widened into a conflict over the construction of a new political hegemony in the city. Berkeley radicals, many of them veterans of the FSM and People's Park, ran their first slate of candidates for city council in 1969, losing to more traditional liberal Democrats who had earlier ousted the conservative establishment that had run the city for the bulk of the 20th century. The radical coalition had much better success in 1973 (and by 1979 had elected one of their own as mayor), and it remained the guiding force in city politics through the 1980s. In combination with liberals, Berkeley radicals reframed Berkeley as a leading center of experimentation for populist-radical politics, including rent control, ecological initiatives, and, until an almost reactionary set of policies was enacted in the 1990s, compassionate care for the homeless and street people (Lyford 1982). Perhaps most importantly, radicals centered on campus and in the South Campus area early aligned themselves with black activists in South and West Berkeley, allowing the activism of the campus to merge (not always easily) with militant black activism in Berkeley and Oakland.

Following the 1969 riots, the university, unable to build dormitories, built a soccer field on a portion of the lot. Students and community members staged a successful boycott of the field, and the university abandoned it a few years later. For a time during the 1970s, a portion of the park reverted back to a free parking lot, but when the university proposed charging fees, the community responded with jackhammers and destroyed the lot in front of onlooking and passive police. In 1976 the university held hearings on developing married student housing on the site. Faced with overwhelming opposition, the university eventually withdrew the plan.

During all these battles of the 1970s and into the 1980s, the park became a growing refuge, not only for political action but also for (mostly male) homeless people. Indeed, the growth of the homeless or transient population, coupled with the vehemence with which park defenders opposed development on the site, led many to see People's Park as a place "off-limits" to students and police alike. As nearby Telegraph Avenue gentrified during the 1980s, many merchants, students, and visitors began to see the park as a zone of danger and trouble rather than a symbol of radical populist politics—or in some forms as a zone of trouble precisely *because* it was a symbol, and the result, of populist-radical politics (Lyford 1982). For many, a direct line could be drawn from the FSM, through the People's Park riots, and to the state of People's Park in the late 1980s when it was often perceived as an uncontrolled and dangerous sore spot in the side of Berkeley and the university. By 1991 things came to a head once again as rioting broke out on the 20th anniversary of the original riots. But that is a story for the next chapter, a chapter which will use the more recent history of People's Park to explore both the legacy and the meaning of the Free Speech Movement and the creation of the park for any putative right to the city. For rights are exactly what are at stake.

NOTES

1. Locational conflict is usually studied in terms of the siting of specific, often noxious, facilities. The literature is large. A useful review and summary can be found in Takahashi (1998). Much of this literature is concerned with questions of NIMBYism (the "not in my backyard" syndrome), which takes a critical attitude toward often parochial concerns of middle- and upper-middle-class homeowners. However, the locational conflict literature—and activists who struggle for or against the siting of specific facilities—sometimes also intersects with concerns over, and literature on, environmental racism, since it is often the case that noxious and dangerous facilities are "dumped" in poor nonwhite neighborhoods. There is a relationship, that is, between NIMBYism and environmental racism, and the politics of this relationship is both fascinating and critically important. See Pulido (2000). In this chapter I will be turning the argument in a different direction, however, by exploring struggles for free speech, a place for the homeless to hang out, and similar issues as spatial struggles over rights.

2. The issue of campus speech areas arose again in the spring of 2002 with a spate of articles and news reports about various student attempts (notably at West Virginia University) to eliminate free speech zoning on their campus.

3. Indeed, one of the interesting but so far unremarked aspects of the recent attempt to zone speech is that it is *conservative* organizations (such as the Foundation for Individual Rights in Education) that are agitating against them, whereas in the 1960s similar institutions were strongly in favor of zoning—if not eliminating—the speech rights of students. Opposition to free speech zones is thus sometimes couched as an assault *on* free speech in exactly the same manner that Supreme Court Justice Antonin Scalia sees bubble zones around abortion clinics (and patients) as an assault on free speech.

4. Kerr is a fascinating figure. A labor economist by training (and one with great sympathy for the radical organizers of the 1930s), Kerr was the protégé of the important Berkeley sociologist Paul Taylor before becoming president of the largest university system in the country. After being fired as president, Kerr has remained in demand as a theorist of higher education in the modern world.

5. Kerr's *Uses of the University* can be productively read against Bill Readings's (1996) more recent, and critical, *University in Ruins*. It is striking just how much of Kerr's vision for the putatively nonideological "multiversity" has come to pass, and just what the costs, in terms of free, noncommodified inquiry has been.

6. This is an exceedingly interesting and in some senses prescient passage in Kerr et al. (1960). The authors go on to predict exactly the zeitgeist of the 1990s that so celebrated business and advertising triumphalism as a form of rebellion: "Along with the bureaucratic conservativism of economic and political life may well go a New Bohemianism in the other aspects of life and partly as a reaction to the confining nature of the productive side of society. There may well come a new search for individuality and a new meaning to liberty. The economic system may be highly ordered and the political system barren ideologically; but the social and recreational and cultural aspects of life diverse and changing" (295). For an analysis of the 1990s zeitgeist, see Frank (2001).

7. Two standard histories cover the rise of the national student left in the 1950s and 1960s: Miller (1994) and Gitlin (1993).

8. A study conducted by the UC Berkeley student government (ASUC) showed that, of 20 schools with enrollments exceeding 8,000, only the University of Arizona had similarly restrictive regulations concerning the location and content of speech on and off campus by students. While three schools reported no political action among students, 16 reported that there were no substantive hindrances to the exercise of political rights. See Heirich and Kaplan (1965, 30). One of the depressing things about working at a university at the dawn of the 21st century is just how strongly the ideology of *in loco parentis* has been revived by campus administrations— often at the urging of students' parents, parents who themselves fought so hard to dismantle it or were prime beneficiaries of its demise in the first place!

9. The United States House Un-American Activities Committee held a series of hearings in San Francisco during 1960. These hearings were well protested by Berkeley students. On the second day of the hearings, after students had been refused entrance to the hearings chamber, San Francisco police "washed" hundreds of demonstrators down the steps of the San Francisco City Hall. Many of those hosed and many of those arrested in the ensuing roundup were students who had participated in civil rights marches in the South. The congruence of experience was not lost on many. The California legislature had its own Un-American Activities Committee that had been active in witch hunts throughout the 1950s and that was keeping a close eye on the growing unrest on the Berkeley campus. For a history of these activities and the students' role in them, see Heirich and Kaplan (1965) and Draper (1966).

10. Rule 17 had been implemented in response to complaints from farmers' and business groups around the state about support for radical union causes emanating from the Berkeley campus. Keep in mind that the president oversees the whole UC system. A chancellor runs each campus. Thus, speakers and other political activities on specific campuses had to be approved at the system level.

11. Lipset's analysis is remarkably similar to that of an earlier UC professor, Carleton Parker, who, as head of the California Commission of Immigra-

tion and Housing in 1914, dismissed the radical action of the Industrial Workers of the World as an infantile, sexually deviant, psychosis. See Parker (1919) and Mitchell (1996a).

12. There are several chronologies of the FSM, most of which are now available at the impressive Free Speech Movement Archive: *http://www.fsm-a.org/*. These chronologies are often slightly inconsistent with one another. The following is pieced together from these accounts and from published chronologies and analyses such as Draper (1966); Editors of the *California Monthly* (1965); Lipset and Wolin (1965). I make no attempt to resolve minor discrepancies definitively, but rather have deferred to the general sense of when something actually happened. The letter from the dean of students was dated September 14, 1964, but was not received by student groups until September 16.

13. All direct quotations in this section are taken from *http://www.fsm-a.org/ stacks/chron_ca_monthly.html#September%2010*, which is an online version of Editors of the *California Monthly* (1965). As is often the case in questions of free speech, the line between "pure speech" and advocacy or incitement is a very thin one. As Dean Towle explained at one point during the controversy (October 28): "A speaker may say, for instance, that there is going to be a picket line at such-and-such a place, and it is a worthy cause and he hopes people will go. But, he cannot say, 'I'll meet you there and we'll picket.' "

14. Sack (1986) has argued that "emptiable space" is crucial to the development of the modern city. By emptying space of conflicting and uncontrolled uses, control over the lives and activities of its (future) users can be asserted by its owners or other powerful institutions. See also Sibley (1995) and Cresswell (1996).

15. As Annette Kolodny (1975, 4) notes, the creation of People's Park also symbolized "another version of what is probably America's oldest and most cherished fantasy: a daily reality of harmony between man and nature based on an experience of the land as essentially female." This is important, but I do not deal with these issues directly in this volume. Rather, I focus on how the park operated—and operates—as a political space, as a symbol of a "liberated" space in the heart of a capitalist city. That said, the issues of gender that Kolodny raises are crucial to this "political operation," as we will see in the next chapter.

16. Of course, one of the park builders' main claims was that the land was not at all the university's. Park builders sought to drive home the point, among other ways, by tracing and publicizing Native American claims to the land, using Indian imagery on posters and leaflets.

The End of Public Space?

People's Park, the Public, and the Right to the City

STRUGGLING OVER PUBLIC SPACE: THE VOLLEYBALL RIOTS

In April 1989 the chancellor of the University of California, sensing a changing political climate reflected in the moderating radicalism of the Berkeley city council and the general complacency of Reagan–Bush era students, raised once again the idea of building dormitories on People's Park. His timing was not good. Park supporters were in the midst of planning a 20th anniversary memorial of the 1969 riots and a celebration of the park, and the chancellor's proposal contributed to a growing sense that the demise of the park was imminent. Gentrification along Telegraph Avenue, further university development in the South Campus neighborhood, and almost nonstop debates over how best to develop the People's Park site all seemed to indicate that the survival of the park as a user-developed and -controlled site was in jeopardy. The cumulating unease and frustration erupted into a full-scale, if rather nostalgia-tinged, riot on May 20, 1989, the 20th anniversary of the original People's Park riots (*Los Angeles Times* 1989a; *New York Times* 1989).

Central to the 1989 riot was the question of homelessness: what right did homeless people have to the city? And in the city, what right did they have to the park? The growing homeless population in the park, and on the streets of Berkeley more generally, raised critical questions in this supposedly liberated and liberal city—questions about housing policy and gentrification, questions about appropriate street behavior, and questions about just whose city Berkeley was—in ways that they never really were raised in the 1960s. During the 1980s a number of Telegraph Avenue merchants, in particular, were growing increasingly vocal in their demands that "something be done" about the homeless. People's Park as a central gathering and living space for the homeless was seen by a growing number of residents, students, and merchants as not so much the solution to alienated living in the city but rather its cause—only now it was the middle class that was being alienated.

Even so, the university found it hard to do much of anything about People's Park. It had become a—if not *the*—primary symbol of the radical uprisings of the 1960s, and (often self-proclaimed) defenders of the park were zealous in their determination that no new structures be built there and that some of the traditional structures—such as the stage and the "free box" (Figure 4.1)—be preserved at all costs.[1] The politics of the park's evolution was often quite complex, with alliances frequently shifting. Despite their general antipathy for the stance of those merchants who wanted to rid the area of the homeless, many park defenders opposed a 1986 Catholic Worker plan to set up a "People's Café"—a trailer and deck that served as a soup kitchen for Berkeley homeless

FIGURE 4.1. The large central grassy area of People's Park, circa 1990. The Free Stage is behind the cluster of people to the left; the Free Box is below the bulletin board to the right. Photograph by author.

people—in the park. The university found itself aligned with these defenders when it sued to block construction of the People's Café, a suit that a Berkeley judge refused to hear, lecturing the university "about wishy-washy liberalism that refuses to take responsibility for the community's problems" (Horn 1989).

Despite its April announcement, by the fall of 1989, the University of California had resigned itself to maintaining the park as an open space, but it had not yet given up hope of controlling the space and developing the park in its own interest. On November 2 of that year, the chancellor announced that the university would lease a portion of the land to the city of Berkeley for a trial basis. The city determined that its portions of the park would be dedicated to user-control; but it also began exploring ways to remove the homeless people who camped there. To aid in this effort, the chancellor pledged $1 million a year for 10 years to the city to help defray the costs of aid to the homeless and other services (Rabinowitz 1989).

The details of the November 1989 accord took more than another year to be ironed out. During its negotiations with the city, the university emphasized that it had every intention of retaining lot 1875-2 as a park but that it wanted it to be a park in which inappropriate persons— "the criminal element," as the university put it (Boudreau 1991, A3)— were removed to make room for students and middle-class residents who, the university argued, had been excluded as People's Park became a haven for "small-time drug dealers, street people, and the homeless."

The park development plan that the city and university eventually settled on seemed innocuous enough (Figure 4.2): the university would lease the west and east ends of the park (for $1 a year) to the city for "community use." The central portion of the park, the big grassy field where many homeless slept and which was the traditional gathering place for rallies, speeches, and concerts, was to be converted by the university into a recreation area featuring volleyball courts, public pathways, public restrooms, and security lighting. In exchange for the lease, the city was to assume "primary responsibility for law enforcement on the premises" (Kahn 1991a, 28). In addition the university and the city were to establish a "Use Standards and Evaluation Advising Committee," which both hoped would help "bring about a much-hoped-for truce, and realization of the place as a park that everyone can enjoy" (Kahn 1991a, 28). While these developments seemed quite ordinary, all agreed that they portended great change. "To be sure," the suburban

U. C. lease agreement (for land where homeless were living) to redevelop park

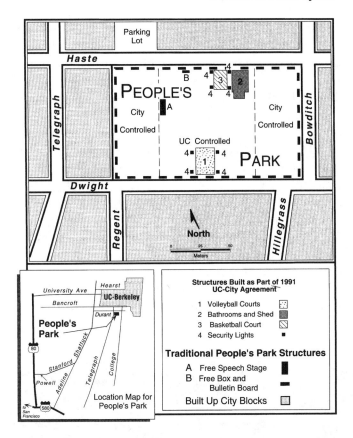

FIGURE 4.2. Map of People's Park showing the developments added in 1991. Map by Jim Robb, University of Colorado.

Contra Costa Times commented, "the one-of-a-kind swath of untamed land will never be the same. And to that extent an era is ending" (Boudreau 1991, A3).

After more than 20 years of riots, debate, controversy, neglect, broken promises, and more riots, the end of the era marked by the city–university agreement seemed long overdue to many in Berkeley and the Bay area. To critics of the park in the city government and university administration, as well as in the mainstream national and local press, the need for improvement in the park was a common theme. Comments such as this were rife in the press at the time: "To some park neighbors and students, People's Park, owned by the university, is overrun by

[handwritten margin note: homeless won't be kicked out, only criminals. Start to associate the homeless as criminals?]

squatters, drug dealers, and the like" (Boudreau 1991, A3) (Figure 4.3). And as the university's director of community affairs, Milton Fujii remarked, "The park is underutilized. Only a small group of people use the park and they are not representative of the community" (*New York Times* 1991a, 1:39). Sensitive to claims that such sentiments might be interpreted as announcing a plan to remove the homeless users of the park, UC spokesperson Jesus Mena declared: "We have no intention to kick out the homeless. They will still be here when the park changes, but without the criminal element that gravitates towards the park" (Boudreau 1991, A3).

For the university, as for other critics of the park, the evident disorder of the park invited criminality. To be a functioning open or public space, it had to be reordered, and the city–university agreement was the first step in that direction. It had to be reclaimed so that it could be made available to an *appropriate* public.

Park defenders saw matters rather differently. For them, People's Park—which after all had survived as a user-controlled space for two

FIGURE 4.3. A homeless encampment in the city-controlled section of the east end of the park, 1994. Photograph by Nora Mitchell; used by permission.

decades despite numerous plans and efforts by the university to reclaim it—constituted one of the few areas in the San Francisco Bay Area in which homeless people could live relatively unmolested (Kahn 1991a, 2) and in which other users had pretty much free reign. It was, especially by contemporary standards, a quite unmediated space. The attractiveness of People's Park for homeless people indicated to many park defenders not that it was a dangerous and out-of-control place but rather that it was working as it should: as a truly public space. It had developed as a political space that encouraged unmediated interaction, a place where the power of the state (and other property owners) could be kept at bay. Activists felt that the university–city accord jeopardized some of the primary park institutions that had developed over the years and that set it apart from city-controlled parks: the grassy assembly area, the Free Speech stage, and the Free Box (a clothes drop-off and exchange) (see Figures 4.1 and 4.3). Without these, they felt that People's Park *as such* would cease to exist. According to Michael Delacour, one of the founders of the park in 1969, the defense of People's Park against the university's plans was "still about free speech, about giving people a place to go and just be, to say whatever they want" (Lynch and Dietz 1991, A20).

By the time the university and city were ready to act on their plans for park development, in the summer of 1991, activists had successfully linked this aspect of the park—the ability for people "to go and just be"—to the rights of homeless people. For those opposed to the UC–city plan, People's Park since its inception had been regarded as a refuge for the homeless and other street people. Activists feared that the building of volleyball courts struck at the heart of the park's traditional role—a place to "just be." As such, it portended ill for homeless park users and residents. It signaled a desire to see their removal. Reconstructing the park in such a way that would lead to the removal of the homeless, they surmised, was tantamount to an erosion of public space. The development of even volleyball courts had to be resisted.

Homeless residents in the park agreed. In her reply to a reporter who asked her about the UC–city plans, Virginia, a homeless woman living in the park, voiced the fears of many homeless people in the park and of park activists: "You know what this is about as well as I do, it is only a matter of time before they start limiting the people able to come here to college kids with an ID." When the reporter reminded her that the university promised not to remove the homeless, Virginia responded: "You look smarter than that. A national monument is being torn down" (Rivlin 1991a, 27). Oakland Homeless Union activist Andrew Jackson put the struggles over People's Park into a larger context. Looking at the bulldozers at work as redevelopment began in the park, he commented: "They're tearing up a dream. . . . Ever since I remember this has been a place for all people, not just for some college kids to play volleyball or the white collar. It's a place to lie down and sleep when you are tired" (Rivlin 1991a, 27).

Activists and homeless residents alike considered changes in the park to be related to changes on nearby Telegraph Avenue—the very avenue that UC Regents had long ago worried about becoming a center for "hippies" and other undesirable counterculture figures. Echoing the arguments made by and against Free Speech activists a generation earlier, People's Park activists feared that a transformed and "tamed" People's Park would become a beachhead for the wholesale transformation of the surrounding neighborhood. "The university says they're not against homeless people," commented homeless activist Curtis Bray soon after the city–university accord was finalized,

> but all the rules and regulations that are coming out for the park are regulations that only affect the homeless community and no one else. . . . They don't want their students to be faced on a daily basis with what it is like to be poor and in poverty. Once they get the cement courts in, they're going to want to keep the homeless population out as much as possible. (Kahn 1991a, 2, 28)

Bray predicted that the agreement on People's Park was just the beginning. "Once People's Park is off-limits, the homeless are going to go to [Telegraph] Avenue. The university will then say the avenue is a problem" (Kahn 1991a, 28). David Nadle, another founder of the park and later an owner of a world-beat dance club in the city,[2] concurred. He denounced the UC–city agreement as a final move toward the total commodification and control of space. "The corporate world is trying to take Berkeley. The park represents a 22-year struggle over corporate expansion."[3] Berkeley, he claimed, had become "yupped out" (Kahn 1991b, 30).

In the years since the 1969 riots, Telegraph Avenue had experienced a series of transformations. A popular gathering place for Bay Area teens, the Telegraph Avenue–People's Park area was both a highly successful commercial district and one always on the brink of decline.

By the early 1980s, the counterculture flavor of the street (with its numerous locally owned coffee houses, bookstores, record shops, and street vendors selling everything from political bumper stickers to drug paraphernalia to locally produced arts and crafts) was beginning to yield to businesses catering to more affluent students and young professionals. By mid-decade, chain stores were beginning to expand at the expense of locally owned businesses. Coffee bars that appealed to the slumming suburban middle class replaced many of the small restaurants and "head" shops that had defined the street throughout the 1970s. Graffiti- and poster-covered walls were replaced with pastel colors and trendy neon.

As the boom times of the 1980s turned into the bust of the early 1990s, many students in the South Campus area, for whom the upheavals of the 1960s were not even a distant memory, had little time or patience for street activism and street spectacle. Both the park and the avenue reflected these changes in political and economic climate. "In a city where protesting was once as common as jogging," wrote the *San Francisco Chronicle* (Lynch and Dietz 1991, A1), "there is little tolerance for uprisings." As park activist Michael Delacour observed, "The students have changed. They know times are tough and they want to survive" (Lynch and Dietz 1991, A20). Time was scarce for activism and the community involvement that makes spaces like People's Park possible. Many students simply avoided the "untamed land" of People's Park. Others students who lived in apartments or dorms neighboring the park strongly supported the university's plans to take control over it.[4]

In the early 1990s, some of the chain stores moved out of Telegraph (Figure 4.4), and an air of dilapidation seemed to settle over the avenue (May 1993, 6) as visible homelessness increased. While many merchants attributed the decline to the physical hazards that People's Park and some of the people who used it posed to middle-class shoppers, officials of the Telegraph Avenue Merchants Association conceded that it was in fact the *image* of the park (and the avenue) that was threatening business success. As one official of the association put it: "If the majority of people think it's unsafe, unclean, why do they think that? Isn't it based on some sort of reality?" (Kahn 1991a, 28). The official did not directly answer her own question (and if she had, she would have had to concede that crime rates were no higher around Telegraph than in other commercial districts of the city). But perhaps, on Thursday, Au-

FIGURE 4.4. The old Miller's Outpost store, circa 1993. Part of a chain of clothing shops, the Miller's Outpost was seen as a symbol of both Telegraph Avenue's gentrification and its decline. Photograph by author.

gust 1, 1991, others did for her, providing the "reality" to which she referred.

On that morning, about 20 activists were arrested as they protested the bulldozers that, in the first step of implementing the UC–city park accord, began clearing grass and soil along the southern edge of the central part of the park where two sand volleyball courts were to be constructed (see Figure 4.2). By that evening, police found themselves trying to control a full-scale riot, as park defenders battled in the streets over whether park development could continue. Intense rioting around the park continued for the next 3 days, and smaller violent conflagrations continued to erupt for almost a week. Police repeatedly fired wood and putty bullets into crowds, and reports of police brutality were widespread (including the witnessed beating of a member of the Berkeley Police Review Commission). But neither did protesters refrain from violence, heaving rocks and bottles filled with urine at the police.[5]

The papers that week were filled with reports of street skirmishes, strategic advances by heavily armed police, and the rage felt by many protesters. Police were accused of beating bystanders, roughing up homeless residents of the park, and using wood and putty bullets needlessly. Police countered that force was necessary to quell the riots, which included numerous street fires, protesters throwing rocks and bottles, smashed windows and vandalism along Telegraph Avenue. By August 6, eight formal complaints of police brutality had been filed with the Police Review Commission and six with the police department itself. A Police Commission member had received 50 statements alleging police abuse, and the commission itself received another 25 calls of complaint. In addition, an unknown number of police were injured in the rioting (Rivlin 1991b, 18).

"We offered to negotiate," club owner David Nadle claimed, "but this is what we got. Militarily, they have commandeered that part of the park"—the center zone with the Free Speech area, the human services, and the Free Box (Kahn 1991c, 11). The police occupation succeeded. The most intense rioting had all but subsided by Saturday, August 3, and park defenders conceded defeat. In a rally of protesters in the park the next day, park founder Michael Delecourt declared: "Basically, we've got no real choice over what happens in this Park anyway" (Auchard 1991, 23)—a remarkable concession after 22 years of tenacious struggle to maintain user control over the area.

Four days later, the first volleyball games were played in the park. Seeking to cement what one Park defender had earlier called "dominion, imposing solutions for other people's own good" (*New York Times* 1991b, A8), university officials released student employees from their jobs provided that they would play volleyball in the park. One of the players, a Berkeley junior and housing office employee, told the *San Francisco Chronicle* (Lynch 1991b, A20):

> At first I thought "OK, let's go play volleyball." But then I realized there was more at stake and I got a little scared. But I came out here because I want to see this happen and show my support. People's Park needs to change. I've only been here once before—most people think the place isn't safe.

That evening at 7 P.M., despite the absence of disturbances around the park or along Telegraph Avenue, police arrested 16 people for trespassing after the park—which the university had asserted it wanted to retain as open space, and from which, the university said, it had no intention of removing the homeless—was closed (Lynch 1991b).

THE DIALECTIC OF PUBLIC SPACE

The Berkeley housing employee was right. There was a lot more at stake in People's Park than volleyball. Most directly, as Duane, a homeless man who lived in the park put it, "This is about homelessness, and job-lessness, and fighting oppression" (Koopman 1991, A13). It was, in other words, about rights, and about the right to the city. But such rights—to a home and job, and to freedom from oppression—were structured through a struggle over a right to and for public space, what such space means, and for whom it is "public." Among other issues at stake in the riots were two opposed and perhaps irreconcilable ideological visions of the nature and purpose of public space, two opposed visions that have a great deal of impact on how the right to the city is conceptualized and for whom it is a viable right. Activists and the homeless people who used the park promoted a vision of a space marked by free interactions, user determination, and the absence of coercion by powerful institutions—in other words, the same sorts of ideological visions for public space promoted by the Free Speech Movement a generation earlier. For them, public space was an unconstrained space within which political movements could organize and expand into wider arenas (see N. Smith 1992a, 1993). The vision of representatives of the university was quite different. Theirs was one of a space that was open for recreation and entertainment, subject to usage by an *appropriate* public (students, middle class residents and visitors, etc.) that used the space by permission of its owners. Public space is imagined in this vision to be a controlled and orderly *retreat* where a properly behaved public might experience the spectacle of the city. In the first of these visions, public space is taken and remade by political actors; it is politicized at its very core; and it tolerates the risk of disorder (including recidivist political movements) as central to its functioning. In the second vision, public space is planned, orderly, and safe. Users of this space must be made to feel comfortable, and they should not be driven away by unsightly homeless people or unsolicited political activity. These visions, of course, are not unique to Berkeley. They are, in fact, the predominant ways of seeing public space in contemporary cities.[6]

   If these two visions of public space indicate that differing definitions of the right to the city are at stake, then they also correspond more or less with Lefebvre's (1991) distinction in *The Production of Space* between *representational space* (appropriated, lived space; space-in-use) and *representations of space* (planned, controlled, ordered space).[7] Public space often, though not always, originates as a representation of space, as for example a courthouse square, a monumental plaza, a public park, or a pedestrian shopping district (Harvey 1993; Hershkovitz 1993; Sorkin 1992). But as people use these spaces, they also become representational spaces, appropriated in use. Public space is thus socially *produced* through its use *as* public space.

In the case of People's Park, however, the standard chronology was in many ways reversed. People's Park began as a representational space, one that had been *taken* and appropriated from the outset.[8] It was wrested from the university (who had already taken it from its previous residents and owners). But whatever the origins of any public space (planned, appropriated, accidental), its status as "public" is created and maintained through the ongoing opposition of visions that have been held, on the one hand, by those who seek order and control and, on the other, by those who seek places for oppositional political activity and unmediated interaction.

If public spaces arise out of a dialectic between representations of space and representational spaces, between the ordered and the appropriated, then they are also, and very importantly, *spaces for representation*. That is, public space is a place within which political movements can stake out the territory that allows them to be *seen* (and heard)—as the IWW understood so well in its struggles for free speech in the city in the first decades of the 20th century (Chapter 2). If the right to the city is a cry and a demand, then it is only a cry that is heard and a demand that has force to the degree that there is a space from and within which this cry and demand is visible. In public space—on street corners or in parks, in the streets during riots and demonstrations—political organizations can represent themselves to a larger population, and through this representation give their cries and their demands some force.[9] By claiming space in public, by creating public spaces, social groups themselves become public.[10] *Only* in public space, for example, can the homeless represent themselves as a legitimate part of "the public." Insofar as homeless people or other marginalized groups remain invisible to society, they fail to be counted as legitimate members of the polity.[11] And in this sense, public spaces are absolutely essential to the functioning of democratic politics (Fraser 1990). Public space is the product of competing ideologies about what constitutes that space—order and control or free, and perhaps dangerous, interaction.[12] These are not merely questions of ideology, of course. They are, rather, questions about the very spaces that make political activities possible. To understand, therefore, why a plan to build volleyball courts and public restrooms in People's Park led to such extreme violence, to understand why people can be so passionate about spaces such as People's Park, we need to reexamine the normative ideals that drive political activity and the nature of the spaces we call "public" in democratic societies. Doing so will make it clear that while many of those who seek to order and control public spaces and who seek to make them spaces of exclusion rather than spaces where the cry and demand for the right to the city is heard (and even promoted), while perhaps being only "little Arnold's," are nonetheless giving voice to a definition of "democracy" that needs to be resisted at every turn.

THE IMPORTANCE OF PUBLIC SPACE
IN DEMOCRATIC SOCIETIES

Public space occupies an important—but contested—ideological position in democratic societies. The Supreme Court, as we have seen (Chapter 2), bases its public forum doctrine on the notion that since "time immemorial" people have used the public spaces of the city—the streets, parks, and squares—as gathering places for "communicating between citizens" and "discussing public questions" (*Hague v. CIO* 1939). But, as we have also seen, just *how* and *where* people are to meet, under what conditions they are to do so, and what they are able to discuss are all themselves points of struggle. The central contradiction at the heart of public space is that it demands a certain disorder and unpredictability to function *as* a democratic public space, and yet democratic theory posits that a certain order and rationality are vital to the success of democratic discourse. In practice, the limits and boundaries of "democracy" seem to be determined as much through transgression—as with the Free Speech Movement's insistence on using the campus, against the will of the university, as a space for political organizing—as through legal or bureaucratic ordering. Public space must therefore be understood as always historically and socially contingent, even as it is politically necessary. Attention needs to be paid to the specific practices through which public space is produced and how the power to determine its use is arrayed.

All that said, it is nonetheless important to sketch, even if only very briefly, the history of public space as both a form and an ideology. The notion of urban public space can be traced back at least to the Greek *agora* and its function as "the place of citizenship, an open space where public affairs and legal disputes were conducted . . . " (Hartley 1992, 29). While the *agora* was thus a political space, "it was also a marketplace, a place of pleasurable jostling where citizens' bodies, words, actions, and produce were all on mutual display, and where judgments, decisions, and bargains were made" (Hartley 1992, 30). Politics, commerce, and spectacle were juxtaposed and intermingled in the public space of the *agora*. It provided a meeting place for strangers, whether citizens, buyers, or sellers, and the ideal of public space in the *agora* encouraged nearly unmediated interaction—the first vision of public space referred to above. In such "open and accessible public spaces and forums," as Iris Marion Young (1990, 119) has put it, "one should expect to encounter and hear from those who are different, whose social perspectives, experience and affiliations are different."[13] One should expect, that is, *urban* experiences, defined by conflicting demands for the right to the city.

Young is speaking specifically of a normative *ideal* of public space. In "actually existing democracies" (Fraser 1990), the functioning of public spaces has rarely lived up to the ideal. The normative ideal that Young points to has its echo in Habermas's (1989) analysis of the (aspatial) normative public *sphere*, which argues that the bourgeois public sphere developed in early modern Europe as the ideal of a suite of institutions and activities that mediated the relationship between the state and society (see Howell 1993; Calhoun 1992). In this normative sense, the public sphere was where "the public" was organized and represented (or imagined). The public sphere is normative, because it is where all manner of social formations should find access to the structures of power within a society (Habermas 1989). Many theorists (e.g., Fraser 1990; Hartley 1992; Howell 1993) contend that public space serves as the material location where social interactions and public activities of all members of "the public" occur. Public space is the space of the public.

Just what that "space" *is*, however, is a point of deep contention. As has already been made clear, the streets and parks of the city, like the Greek *agora*, Roman *forums*, or 18th-century German coffeehouses (Habermas 1989) before them, have never simply been places of free, unmediated interaction. Rather, they have always also been spaces of exclusion (Fraser 1990; Hartley 1992). The public that met in these spaces was carefully selected and homogeneous in composition (contra Young's ideal). It consisted of those with power, legal standing, and respectability, and in this exclusiveness the roots of the second vision of public space can be seen. In Greek democracy, for example, citizenship was a right denied to slaves, women, and foreigners. None of them had standing in the public spaces of Greek cities, even as their labor (and their money) may have been welcomed in the *agora*. They were formally excluded from the political activities of the public space.

And in American history, of course, the admittance of women, the propertyless, and people of color into the formal ranks of "the public" has been startlingly recent (and not yet really complete). Foreigners are still not considered part of the public (and recent changes in immigration law have in fact eroded rights for many noncitizen residents). Women, some of the propertyless, and people of color have only won entrance to "the public" through concerted social struggle, demanding the right to be seen, to be heard, and to directly influence the state and society. As in other Western countries, notions of "the public" and the nature of public democracy played off and developed dialectically with both the fact and the ideology of private property and the private sphere. The ability for citizens to move between private property and public space determined the nature of public interaction in the developing democracy of the United States (Fraser 1990; Habermas 1989; Marston 1990). In the context of an evolving capitalist American state, citizenship is defined through a process whereby "owners of private property freely join together to create a public, which forms the critical functional element of the public realm" (Marston 1990, 445). To be public means having access to private space to retreat to (so that publicness can remain voluntary).

Each of these spheres—the public and the private—of course has been constrained and defined by gender, class, and race. By the end of the 18th century, according to Richard Sennett (1992, 18–19, emphasis in the original):

> The line drawn between public and private was essentially one on which the claims of civility—epitomized by cosmopolitan, public behavior— were balanced against the claims of nature—epitomized by the family. . . . [W]hile man made himself in public, he *realized* his nature in the private realm, above all in his experiences within the family.

The private sphere was the home and refuge, the place from which white propertied men ventured out into the democratic arena of public space.[14] The public sphere of America and other capitalist democracies was thus understood as a voluntary community of private (and usually propertied) citizens. By "nature" (though really by custom, economics, franchise, law, and sometimes outright force) women, nonwhite men, and the propertyless were denied access to the public sphere in everyday life.[15] Built on exclusions, the public sphere was thus a "profoundly problematic construction" (Marston 1990, 457).

For the historian Edmund Morgan (1988, 15), the popular sovereignty that arose from the split between publicity and privacy was a fiction in which citizens "*willingly* suspended disbelief" as to the improbability of a total public sphere.[16] The normative idea of the public sphere holds out the hope that a *representative* public can meet (Hartley 1992). The reality of public space and the public sphere is that Morgan's "fiction" is less an agreeable acquiescence to representation and more "an exercise in ideological construction with respect to who belongs to the national community and the relationship of 'the people' to formal government" (Marston 1990, 450). It is precisely a contest over who counts as Morgan's "citizens" (see Brown 1997).

As ideological constructions, contested ideals such as "the public," public space, and the public sphere take on double importance. Their very articulation implies a notion of inclusiveness that becomes a rallying point for successive waves of political activity. Over time, such political activity has broadened definitions of "the public." It is no longer so easy (though still possible) to exclude women, people of color, and some of the propertyless from a formal voice in the affairs of state and society. In turn, redefinitions of citizenship accomplished through struggles for inclusion have reinforced and even transformed normative ideals incorporated in notions of the public sphere and public space. By calling on the rhetoric of inclusion and interaction that the public sphere and public space are meant to represent, excluded groups have been able to argue for their *rights* as part of the active public—to make a claim for a right to the city. And each (partially) successful battle for inclusion in "the public" conveys to other marginalized groups the importance of the ideal as a point of political struggle (even as it also calls opponents of widening "the public" to the barricades, or at least to the lofty pulpits of the right-wing think tanks).

In these struggles for inclusion, the distinctions between the public sphere and public space assume considerable importance. The public sphere in the sense that Habermas developed it and many of his critics have refined it is a universal, abstract sphere in which democracy occurs. The materiality of this sphere is, so to speak, immaterial to its functioning. Public space, meanwhile, is material. It constitutes an actual site, a place, a ground within and from which political activity flows.[17] This distinction is crucial, for it is "in the context of real public spaces" that alternative movements may arise and contest issues of citizenship and democracy (Howell 1993, 318).

If contemporary trends signal an erosion of the first vision of public space as the second becomes more prominent (see below; Crilley 1993; Davis 1990; Fyfe 1998; Gold and Revill 2000; Goss 1992, 1993; Sennett 1992; Sorkin 1992), then spaces such as People's Park become, in Arendt's words, "small hidden islands of freedom" (quoted in Howell 1993, 313).[18] Such hidden islands are created when marginalized groups *take* space and use it to press their claims, to cry out for their rights. And that was precisely how activists understood their defense of People's Park in the face of the university's desire to transform and better control it. As the *East Bay Express* observed (Kahn 1991c, 11): "Ultimately, they claim, this is still a fight over territory. It is not just two volleyball courts; it's the whole issue of who has a rightful claim to the land." Michael Delacour argued that People's Park was still about free speech, and the homeless activist Curtis Bray claimed that "they are trying to take the power away from the people" (*New York Times* 1991a, 1:39). For these activists, People's Park was a place where the rights of citizenship could be expanded to the most disenfranchised segment of contemporary American democracy: the homeless. People's Park provided the space for representing the legitimacy of homeless people within "the public." In just this sense, People's Park *was* exactly that sort of Hyde Park that Matthew Arnold railed against. Like the streets of San Diego for the IWW 80 years earlier, People's Park was, for homeless people, a deeply *political* space.

THE POSITION OF THE HOMELESS
IN PUBLIC SPACE AND AS PART OF THE PUBLIC

People's Park has been recognized as a refuge for homeless people since its founding, even as elsewhere in Berkeley the city has actively removed squatters and homeless people (sometimes rehousing them in a disused city landfill) and become one of the leading innovators of punitive anti-homeless laws (Dorgan 1985, B12; Harris 1988, B12; Levine 1987, C1; *Los Angeles Times* 1988, 13; Stern 1987, D10; Wells 1994, A14). Consequently, the park has become a relatively safe place for the homeless to congregate—one of the few such spots in an increasingly hostile Bay Area (*Los Angeles Times* 1990, A1). Around the Bay, the homeless have been repeatedly cleaned out of San Francisco's United Nations Plaza near City Hall, Golden Gate Park, and other public gathering places; in Oakland, loitering is actively discouraged in most parks (*Los Angeles Times* 1989b, 13; 1990, A1; MacDonald 1995; *New York Times* 1988b, A14).

In part, the desire to sweep the homeless from visibility responds to the central contradiction of homelessness in a democracy composed of private individuals and private property (see Deutsche 1992; Mair 1986; Marcuse 1988; Ruddick 1990; N. Smith 1989; Takahashi 1998; Waldron 1991). This contradiction turns on publicity: the homeless are all too visible. Although homeless people are nearly always in public, they are rarely counted as part of *the* public. Homeless people are in a double bind. For them, socially legitimated private space does not exist, and so they are denied access to public space and public activity by the laws of a capitalist society that is anchored in private property and privacy (Waldron 1991; Blomley 1994a, 1998, 2000a). For those who are *always* in the public, private activities must necessarily be carried out publicly.[19] When public space thus becomes a place of seemingly illegitimate behavior, our notions of what public space is supposed to be are thrown into doubt. Now less a location for the "pleasurable jostling of bodies" and the political discourse imagined as the appropriate activities of public space in a democracy, public parks and streets begin to take on aspects of the home. They become places to go to the bathroom, sleep, drink, or make love—all socially legitimate activities when done in private but seemingly illegitimate when carried out in public (Staeheli 1996).

As importantly, since citizenship in modern democracy (at least ideologically) rests on a foundation of *voluntary* association, and since homeless people are *involuntarily* public, homeless people cannot be, by definition, legitimate citizens.[20] In consequence, homeless people have proven threatening to the exercise of rights since they seem to threaten to expose the existence of the "legitimate"—that is, voluntary—public as a contradiction if not a fraud: voluntariness is impossible if some are necessarily excluded from the *option* of joining in or not.

The existence of homeless people in public thus undermines one of the guiding fictions of democracy. This is why George Will (1987) is adamant when he argues (as we saw in Chapter 1) that "Society needs order, and hence has a right to a minimally civilized ambiance in public spaces. Regarding the homeless, this is not merely for aesthetic reasons because the unaesthetic is not merely unappealing. It presents a spectacle of disorder and decay that becomes contagion." The ideological foundation of modern democracy, with all its practical contradictions, is apparently rather fragile. For reasons of order, then, the homeless are continually pushed out of public space, and they are excluded from most definitions of the legitimate public (notice, in Will's formulation, how there is simply no consideration of the *rights* of homeless people as citizens). In much writing about order and the city, the homeless have become something of an "indicator species," diagnostic of the presumed ill health of public space and of the need to gain control, to privatize, or to otherwise rationalize public space in urban places.[21] Whether in New York City (N. Smith 1989, 1992a, 1992b; Zukin 1995), Columbus (Mair 1986), Los Angeles (Ruddick 1996; Takahashi 1998), or Berkeley, the presence of homeless people in public spaces suggests to many an irrational and uncontrolled society in which appropriate distinctions between public and private behavior are muddled (see Cresswell 1996). Hence, those who are intent on rationalizing "public" space have *necessarily* sought to remove the homeless—to banish them to the interstices or margins of civic space, or to push them out altogether—in order to make room for "legitimate" public activities (Mair 1986; Marcuse 1988; Lefebvre 1991, 373).

When, as in Berkeley's People's Park, New York's Tompkins Square and Bryant parks, or San Francisco's Golden Gate Park (Karacas 2000), actions are taken against park users by closing public space or exercising greater social control over park space, the press explains these actions by saying that "the park is currently a haven for drug users and the homeless" (*Los Angeles Times* 1991b, A10; see also Boudreau 1991, A3; Koopman 1991, A13; *Los Angeles Times* 1991a, A3; 1992, A3; *New York Times* 1988a, A31). Such statements, besides creating what are often invidious associations, pointedly ignore any "public" standing that homeless people may have, just as they ignore the possibility that homeless people's usage of a park for political, social, economic, and residential purposes may constitute for them a legitimate and even necessary use of public space. When UC officials claimed that the homeless residents of People's Park were not "representative of the community"

(Boudreau 1991, A3), they in essence denied social legitimacy to homeless people and their (perhaps necessary) behaviors. By transforming the park, UC hoped that illegitimate activity would be discouraged. That is to say that the homeless could stay as long as they behaved "appropriately"—and as long as the historical, normative, ideological boundary between public and private was well patrolled. But that boundary is *itself* a product of constant struggle—especially now in the contemporary city where the neoliberal assault on all things public is in full swing.

PUBLIC SPACE IN THE CONTEMPORARY CITY

Public space is more than just a "Hyde Park"—as crucial as that function is. It is also a representation of the *good* that comes from *public* control and ownership, as contested and problematic as these may be. This is a corollary of the vision of public space as a place of relatively unmediated interaction: it is a vision of public space that understands a space's very publicness as a good in and of itself, that understands there to be a *collective* right to the city. And this vision and practice of public space is increasingly threatened in the American city (as the defenders of People's Park recognized). The threat here is not from the disorderly behaviors of homeless people, as so many argue, but rather from the steady erosion of the ideal of the public, of the collective, and the steady promotion of private, rather than democratic, control of space as the solution to perceived social problems.

The public space of the modern city has always been a hybrid, and certainly a contradictory, space. It is a hybrid of commerce and politics (Sennett 1992, 21–22) in which, ideally at least, the anarchy of the market meets the anarchy of politics to create an interactive, democratic public. In the 20th century, however, markets have increasingly been severed from politics, with, ironically, the latter being banished, fairly completely, from public space. The very success of struggles for inclusion—by women, African Americans, gays, and the propertyless—has led to a strong backlash that has sought to reconfigure urban public space in such a way as to limit the threat of democratic social power to dominant social and economic interests (Fraser 1990; Harvey 1992).

These trends have led to the constriction of public space, even as various social movements continue to struggle for its expansion. Interactive, discursive politics has effectively been banned from the natural gathering places in the city. Corporate and state planners have created environments that are based on a desire for security more than interaction, for entertainment more than (perhaps divisive) politics (Crilley 1993; Garreau 1991; Goss 1992, 1993, 1996; Sorkin 1992; Zukin 1995). One of the results of contemporary urban planning (especially in the post-World War II period) has been the growth of what Sennett (1992) calls "dead public spaces," such as the barren plazas that surround so many modern office towers. A second result, one that evolved as a partial response to the failure of dead public spaces, has been the development of festive spaces that encourage consumption—downtown or seaside festival marketplaces, gentrified historic districts, and even a certain kind of mall (Figure 4.5). Though seemingly so different, both "dead" and "festive" spaces are premised on a perceived need for order, surveillance, and control over the behavior of the public (see Fyfe 1998). As Goss (1993, 29–30) has remarked, we—as consumers and as users of public spaces—are often complicit in the severing of market and political functions. He points to the case of what he calls the "pseudo-public" spaces of the contemporary shopping mall:

> Some of us are . . . disquieted by the constant reminders of surveillance in the sweep of cameras and the patrols of security personnel [in malls]. Yet those of us for whom it is designed are willing to suspend the privileges of public urban space to its relative benevolent authority, for our desire is such that we will readily accept nostalgia as a substitute for experience, absence for presence, and representation for authenticity. (see also Fyfe and Bannister 1995, 1998; Oc and Tiesdell 2000; Williams, Johnstone, and Goodwin 2000)

Goss (1993, 28) calls this nostalgic desire for the market "agoraphilia"— a yearning for "an immediate relationship between producer and consumer."[22]

Such nostalgia is rarely innocent, however (see Lowenthal 1985). It is, rather, a highly constructed, corporatized image of a market quite unlike the idealization of the *agora* as a place of commerce *and* politics (Hartley 1992). In the name of comfort, safety, and profit, political activity is replaced in spaces like the mall, festival marketplace, or redesigned park (such as New York's Bryant) by a highly commodified spectacle designed to sell—to sell either goods or the city as a whole (Boyer 1992; Crawford 1992; Garreau 1991, 48–52; Goss 1996; Mitchell and

FIGURE 4.5. Horton Plaza in San Diego. An example of the playful "festival market" type of privatized public space that has become so important to downtown redevelopment. Photograph by Susan Millar; used by permission.

Van Deusen 2002; Zukin 1995). Planners of pseudopublic spaces such as malls, corporate plazas and redeveloped parks have found that control-led *diversity* is more profitable than the promotion of unconstrained social *differences* (in the sense that Iris Marion Young uses the term) (Crawford 1992; Goss 1993, 1999; Kowinski 1985; A. Wilson 1992; Young 1990; Zukin 1991). Hence, even as new groups are claiming greater access to the rights of society, the homogenization of "the public" continues apace, since the sort of diversity that pseudopublic spaces encourage is a diversity bound up in the unifying, leveling, homogenizing forces of commodity, brand-oriented consumption (Klein 1999).

This homogenization typically has advanced by "disneyfying" space and place—creating landscapes in which every interaction is carefully

planned (Sorkin 1992; A. Wilson 1992; Zukin 1991), right down to specifically planning the sorts of "surprises" one is supposed to encounter
in urban space. Market and design considerations thus displace the id-
iosyncratic and extemporaneous interactions of engaged people in the
determination of the shape of urban space in the contemporary world.
Representations of space come to dominate representational spaces
(Lefebvre 1991; Crilley 1993, 137; Zukin 1991). Designed and con-
trived diversity creates marketable landscapes, as opposed to unscripted
social interaction, which creates places that may sometimes threaten exchange value. The "disneyfication" of space consequently implies the
increasing alienation of people from the possibilities of unmediated social interaction and increasing control by powerful economic and social
actors over the production and use of space.

Imposing limits and controls on spatial interaction has been one of
the principle aims of urban corporate planners during this century (Davis 1990; Fyfe 1998; Gold and Revill 2000; Harvey 1989; Lefebvre
1991). The territorial segregation created through the expression of
social *difference* has increasingly been replaced by a celebration of constrained *diversity*. The diversity represented in shopping centers,
"megastructures," corporate plazas, and (increasingly) public parks is
carefully constructed (Boyer 1992).[23] Moreover, the expansion of a
planning and marketing ethos into all manner of public gathering
places has created a "space of social practice" that sorts and divides social groups (Lefebvre 1991, 375) according to the dictates of comfort
and order rather than those of political struggle. But, as Lefebvre (1991,
375) suggests, this is no accident. The strategies of urban and corporate
planners, he claims, classify and "distribute various social strata and
classes (other than the one that exercises hegemony) across the available territory, keeping them separate and prohibiting all contacts—
these being replaced by *signs* (or images) of contact."[24]

This reliance on images and signs—or representations—entails the
recognitions that a "public" that cannot exist as such is continually
made to exist in the pictures of democracy we carry in our heads: "The
public in its entirety has never met at all . . . "; yet, "the public [is] still
to be found, large as life, in the media" (Hartley 1992, 1). Hence,
"[c]ontemporary politics is *representative* in both senses of the term; citizens are represented by a chosen few, and politics is represented to the
public via the various media of communication. Representative political
space is literally made of pictures—they *constitute* the public domain"

(Hartley 1992, 35, emphasis in original). I will return to the importance of symbolic politics and the resistance it calls up in a moment; for now it is sufficient to note that the politics of symbolism, imaging, and representation increasingly stand in the stead of a democratic *ideal* of direct, less mediated, social interaction in public spaces. In other words, contemporary designers of urban "public" space increasingly accept signs and images of contact as more natural and desirable than contact itself.

Public and pseudopublic spaces perform a vital role in representational politics. The overriding purpose of public space becomes the creation of a "public realm deliberately shaped as theater" (Crilley 1993, 153; see also Glazer 1992). "Significantly, it is a theater in which a pacified public basks in the grandeur of a carefully orchestrated corporate spectacle" (Crilley 1993, 147).[25] This is the purpose of the carefully controlled "public" spaces such as the corporate plazas, library grounds, and suburban streets critiqued by Davis in his important *City of Quartz* (1990, 223–263) and the festival marketplaces, theme parks, historical districts, and so forth analyzed by the contributors to Sorkin's landmark *Variations on a Theme Park* (1992). It is certainly the goal of mall builders (Garreau 1991; Goss 1993; Kowinski 1985; A. Wilson 1992).

These spaces of controlled spectacle narrow the list of people eligible to form "the public." Public spaces of spectacle, theater, and consumption create images that define the public, and these images—backed by law—exclude as "undesirable" the homeless and the political activist. Thus excluded from these public and pseudopublic spaces, their legitimacy as members of the public is put in doubt. And thus *unrepresented* in our images of "the public," they are banished to a realm outside of politics because they are banished from the gathering places of the city.

How "the public" is defined and imaged (as a space, as a social entity, and as an ideal) is quite important. As Crilley (1993, 153) shows corporate producers of space tend to define the public as passive, receptive, and "refined." They foster the "illusion of a homogenized public" by filtering out "the social heterogeneity of the crowd, [and] substituting in its place a flawless fabric of white middle class work, play, and consumption . . . with minimal exposure to the horrifying level of homelessness and racialized poverty that characterizes [the] street environment" (Crilley 1993, 154). And, by blurring the distinctions between private property and public space, they create a public that is narrowly prescribed. The deliberate blurring of carefully controlled spaces (such as Disneyland, Boston's Fanueil Hall, or New York's World Financial Center) with notions of public space "conspires to hide from us the widespread privatization of the public realm and its reduction to the status of commodity" (Crilley 1993, 153). The irony, of course, is that this privatization of public space is lauded by all levels of government (e.g., through public–private redevelopment partnerships) at the same time that the privatization of public space by homeless people (their use of public space for what we consider to be private activities) is excoriated by urban planners, politicians, and social critics alike.

THE END OF PUBLIC SPACE?

The Rise of Open Space

Have we reached, then, "the end of public space" (Sorkin 1992)? Has the dual (though so different) privatization of public space by capital and by homeless people created a world in which designed diversity has so thoroughly replaced the free interaction of strangers that the ideal of an unmediated *political* public space is wholly unrealistic? Have we created a society that expects and desires only private interactions, private communications, and private politics, that reserves public spaces solely for commodified recreation and spectacle? Many cultural critics on the left believe so, as do such mainstream commentators as Garreau (1991) and such conservatives as Glazer (1992). Public spaces are, for these writers, an artifact of a past age, an age of different sensibilities and different ideas about public order and safety, when public spaces were stable, well-defined, and accessible to all. As we have already seen, such sensibilities are nothing more than nostalgic fantasy: the public spaces of the past were anything but inclusive, except insofar as concerted social protest and conflict opened them up for new groups of people and new kinds of politics. That is, then as now, public spaces were only "public" to the degree that they were *taken* and made public. Definitions of public space and "the public" are not universal and enduring; they are produced through constant struggle in the past and in the present. In People's Park, as in so many other places (such as the streets of Seattle and Washington during ministerial meetings, or over the fight to preserve community gardens in New York), that struggle continues.

But that said, the places where the struggle may open up—that is, the opportunities for *taking* space—are steadily diminishing as new forms of surveillance and control are implemented (even though many cities are in fact increasing their stock of open spaces and parklands). During the period of rapid suburbanization and urban renewal in the decades after World War II, North American cities "vastly increased open space, but its primary purpose was different [than public spaces with civic functions], i.e., to separate functions, open up distance between buildings, allow for the penetration of sunlight and greenery, not to provide extensive social contact" (Greenberg 1990, 324).[26] There are many reasons for the growth of open space: preserving ecologically sensitive areas, maintaining property values by establishing an undevelopable greenbelt, providing places for recreation, removing flood plains from development, and so on. But in each case open space serves functional and ideological roles that differ from political public spaces. Indeed, open spaces often share characteristics with pseudopublic spaces: restrictions on behavior and activities are taken for granted; prominent signs designate appropriate uses and outline rules governing where one may walk, ride, or gather. These are highly regulated spaces.[27]

In Berkeley, UC officials recognized this distinction between open and public space. During various People's Park debates, speakers for the university never referred to the park as public space (even though the land is owned by a public entity), though they frequently reiterated their commitment to maintaining the park as open space (Boudreau 1991, A3). Berkeley City Council member Alan Goldfarb, an occasional critic of university plans, also traded on the differences between public and open space. Speaking of People's Park, he celebrated the virtues of public space and then undermined them:

> It's a symbol for the police versus the homeless, the have-nots versus the haves, progress versus turmoil, all the undercurrents most troubling in the city. You've got pan-handling going on, the business community nearby, the town–gown tensions. You have anarchists and traditionalists. People's Park becomes a live stage for all these actors. For many people around the world, Berkeley is People's Park. (Kahn 1991a, 28, emphasis in original)

But if "[t]hese things are real and important," he continued, it is even more important to make People's Park "a viable open space" that would provide a bit of green in a highly urbanized area (Kahn 1991a, 28). The end of public space might consist as much in its deracination as in its co-optation by corporate or state interests.

New Public Spaces?

But there is an even stronger argument for the end of public space than its usurpation by a suburban ideal of open space. Many analysts suggest that the very nature of space has been transformed by developments in communications technology—even to the point where the right to the *material* city is decreasingly necessary, so long as one has access to the "city of bits" (W. Mitchell 1995). They maintain that the electronic space of the media and computer networks has opened a new frontier of public space in which the material public spaces in the city are super-seded by the forums of (perhaps interactive) television, talk radio, and the web. For many scholars (not to mention all those entrepreneurs who rode the dot-com wave to untold riches—at least until the wave crashed on rocky shores) modern communications technology now provides the primary site for discursive public activity in general and in politics in particular. Indeed, such a sense was all-pervasive, perhaps best gauged by the desire of newspapers to ever more closely track the pulse of "the people" by printing transcripts of "what they're saying on the web." Recall how not a political or social event of the mid- to late 1990s—Princess Diana's death, the Monica scandal, even the various crises in Kosovo and Belgrade—could pass without every newspaper in the land tuning into the various chat rooms so they could track for those of us less well connected just what "we" were thinking.[28] And, in-deed, there was something of an explosion of discursive populism spurred by the web and talk radio and TV. But defining chat rooms, fax broadcasts, talk radio, and television as "public space" is not an unproblematic move, even if the media (newspapers and so forth) have always been bound up in the construction of national "publics" (Ander-son 1991; Habermas 1989). If we have indeed created "the first cyberspace nation" (Roberts 1994, C1), then our very conceptions of citizenship have been transformed without much by way of public de-bate—or much by way of the struggle for inclusion that typically marks such transformations. One might immediately want to ask who has been excluded in this move. One might also want to know what it means when being part of "the public" no longer requires being *in* the public, but instead can be accomplished from the private home by tuning the radio, switching the TV channel, or dialing up the modem. And yet these questions seem rarely to be raised.

Consider, for example, the rather optimistic account of electronic space as public space by the Mass Media Group (MMG) of the Committee for Cultural Studies at CUNY Graduate School. Writing before the explosion of the web and focusing on television, the MMG challenged the second part of what they deemed the "unquestionable truism" that "the media today *is* the public sphere, and this the reason for the degradation of public life if not its disappearance" (Carpignano et al. 1990, 33, emphasis in original). The MMG argues instead that the evolution of television talk shows has transformed "the public" from an audience for mass politics and entertainment into a discursive interactive entity—a proto-web, perhaps. TV talk shows "constitute a 'contested space' in which new discursive practices are developed in contrast to the traditional modes of political and ideological representation" (Carpignano et al. 1990, 35).

For the MMG, talk shows are now "common places" that produce "common sense" in a manner analogous to idealized town meetings of times past: "Common sense could also be defined [within these shows] as the product of an electronically defined common place which, by virtue of being electronically reproduced, can be considered a public space. In its most elementary form, going public today means going on the air" (Carpignano et al. 1990, 50). MTV put it even more bluntly after the 1992 presidential campaign (and reprising this, too, for the 1996 and 2000 campaigns). On November 9, 1992, the network ran full-page ads in newspapers across the nation "salut[ing] the 17 million 18–29 year olds who stood up, turned out and voted." The advertisements carried the logo "MTV, the community of the future." As with MTV's vote drive campaign as a whole, the ads were "presented by AT&T, The Ford Motor Company, and your local cable company." MTV's campaign tempers the MMG's optimistic assessment of the power of the electronic media "in the age of chatter": corporate sponsorship, MTV makes clear, is what makes public space possible.[29] The similarities between what the MMG hails as the "therapeutic" discursive practices of the talk show (Carpignano et al. 1990, 51; see also Sennett 1992, 12, 269–293) and the privatization and corporate control of public space are readily apparent. In both cases, the material structure of the medium closes off political possibilities and opportunities. The "public" gathering in the "public space" of the afternoon talk show (contra the MMG's claim that it is unmediated) is a selected audience that is scripted *in advance*.[30] Members of the audience are expected to be articulate, to stake out controversial positions, and to add to the spectacle while at the *same time* not alienating sponsors or viewers. MTV's structuring of the community of the future, along with the MMG's assessment of contemporary public space, provides ample evidence that the sorts of commodification of public space that are so apparent in the material realm are also well advanced in the electronic.

As importantly, if it is true that "going public . . . means going on the air," this undermines the use of *material* public spaces for democratic politics. If the MMG is correct, then politics will henceforth *only* be possible through the media, only through highly structured and dominated electronic "spaces." The MMG puts the best face on this situation by suggesting that the nature of the talk-show format, its compromise between confrontation and shock, "becomes an opening for the empowerment of an alternative discursive practice" (Carpignano et al. 1990, 52). Yet, this empowerment is almost exclusively a private, solipsistic empowerment of therapy, and one which has little to say about alternative political projects.[31] Television chat shows, like "disneyfied" city spaces, create a certain kind of "public"—one in which individuals are allowed to get angry, albeit in their place and in a highly scripted manner, but one that is ultimately nonthreatening to established structures of order and power. The spectacle of "the public" is dissolved into public spectacle.

But TV is old media; it is discredited one-to-many broadcasting.[32] Surely the one-to-one, one-to-many, and many-to-one capabilities of the Internet allow the best chance yet to realize the ideal of unmediated interaction and communication embodied in some visions of public space. When I first started writing about People's Park and public space, the Internet revolution was just emerging, but even then some of the directions in which the revolution was going to be pushed were clear. The federal government worked assiduously to privatize most of the infrastructure during the early 1990s, suggesting that the Internet was not to be seen as primarily a *public* good. But what was hard to predict was just how quickly and how thoroughly the "space" of the web would be commercialized, essentially commodifying vaunted one-to-one communication in ways it never had been before. There is no doubt that the net has been a vital force for political organizing. It has become an indispensable means of communicating between activists and activist groups. The groundswell of prodemocratic protests that have greeted every major economic and political summit during the past few years might not have had the force—and surely could not have organized the sheer numbers and varieties of people—that they have without the communicative capabilities of the web. But in this sense the role of the web is to serve the same function as the telephone and the newsletter used to— only much more efficiently and in a way that allows close to real-time communication across vast distances. This is important, but what is more important were the people—their bodies and their costumes, even their rocks and bottles—on the city streets. It was their *visibility* in the *material* public spaces of the summit cities that has made the difference. All the web communications in the world would not have nearly shut down the Seattle meeting of the World Trade Organization or destroyed the Genoa talks. But people in the streets did.

What is remarkable about the web, to put all this another way, is just how little public visibility it has. Indeed, its main function is to facilitate private (or small-group) communication, to make more efficient the publishing of newsletters, magazines, and tracts (which can all now reach a larger potential audience), and to act as a giant catalog showroom. Just as importantly, electronic communication embodies a rather different ideal of public space than that of the *agora* (despite the promiscuous mixing of market and politics that is so much a part of the web), and it responds to a different set of social desires. "What society expects, and [cyberspace] exemplifies, is to conduct itself via a private ethic of transmissive communication" (Hillis 1994, 191), and the web is becoming the perfect technology for this desire. Such a desire, and its fulfillment, however, is remarkably limited and diminishing: as Setha Low (2000, 247) notes, "in cyberspace we cannot see, hear, touch, and feel each other, much less our environment." And, of course, in cyberspace, we cannot *live*. A fully electronic public space renders marginalized groups such as the homeless even more invisible to the workings of politics (Hillis 1994): there is literally no room in the Internet's "public space" for a homeless person to exist—to sleep, to relax, to attend to bodily needs. Nor can the needs, desires, and political representations of the homeless ever be *seen* in the manner that they can be seen in the public spaces of the city. It is a limited political world, indeed, that assumes that only those who *can* "go on the air" *need* to "go public" with their representations.

THE NECESSITY OF MATERIAL PUBLIC SPACES

The vision of the electronic future as public space has proven, by any number of events—from the uprisings in Tiananmen Square, Leipzig,

Prague, and Budapest in 1989, to the anti-corporate globalization protests in Seattle, Bangkok, Quebec City, Davos, and Genoa at the dawn of
the new century, to the growing "take back the streets" movements in
countless cities in Europe and North America—to be little more than
wishful thinking. It has proven, despite the importance of electronic
communications for *organizing*, to be more a dream of control than
liberatory democracy. This is so simply because public democracy requires public visibility, and public visibility requires material public
spaces. This is not to say that electronic media are not important—quite
the contrary—but it is to say that they are not even close to sufficient.

Consider the uprising in Tiananmen Square. Electronic communications—the telephone and fax in particular—played an important role
in organizing the protest, but the uprising truly began only with the
transformation of the square itself from a monumental and official space
(a space of representation) "into a genuine place of political discourse"
(a representational space) (Calhoun 1989, 57). Students, workers, and
other activists "met in small groups of friends, large audiences for
speeches, and even more or less representative council for debating
their collective strategy and carrying out self-government" (Calhoun
1989, 57). But the important thing is that this *mass* movement took
over a specific—and centrally important—physical *space*. As Hershkovitz (1993, 417) suggests, the appropriation of Tiananmen served as
incisive "evidence of the extraordinary power of apparently 'placeless'
movements to create and transform space in new and authentically revolutionary ways."[33] By taking over and transforming the square, the
movement created a space *for* representation—representations that were
then picked up by the media and broadcast around the world. Without
capturing the space of the square (and, indeed, without being incredibly
savvy in timing the protest), the movement simply would not have been
seen—at least not at the scale, and with the impact, that it was.

Spaces such as Tiananmen Square (or the central square in Leipzig)
enable opposition to be extended to wider scales, to radiate out into the
wider polity. This is no less true of People's Park, even if the events there
may not have had the immediate world-historical importance of the
events in Leipzig and Beijing. *After* space is taken—whether that space
is a contested city lot, as in Berkeley, or the most important public space
in China—oppositional representations expand beyond the confines of
the local struggle, in part because they are broadcast (D'Arcus 2001).
Without the occupation of the space, without taking it, however, the

kinds of protests that came to a climax in Tiananmen, Leipzig, Seattle, or People's Park would have remained invisible. The occupation of space is a necessary ingredient of protest, a fact that the forces of the state, capital, or other powers know only too well.

For this reason, reliance on the media as the entrée into the public sphere is dangerous (Fraser 1990). Media in the "bourgeois public sphere" (as it has been described by Habermas) "are privately owned and operated for profit. Consequently, subordinated social groups lack equal access to the material means of equal participation" (Fraser 1990, 64–65). To overcome the problem of access, "subaltern counter publics" create a "parallel discursive arena where members of subordinate groups invent and circulate counterdiscourses, which in turn permit them to formulate oppositional interpretations of their identities, interests and needs" (Fraser 1990, 67). In these arenas and spaces, counterpublics can be seen by other factions of the public. Without these spaces, "the public" is balkanized. Occupation of public space, then, "militates in the long run against separatism because it assumes an orientation that is *publicist*. Insofar as these arenas are *publics* they are by definition not enclaves—which is not to say that they are not often involuntarily enclaved" (Fraser 1990, 67, emphasis in original). This is exactly the dynamic at work in the current round of anti-corporate globalization protests, even if many of the condescending tribunes of the global elite (such as Thomas Friedman or Paul Krugman at the *New York Times*) fail to understand that.

While television and other electronic media (including the Internet) have important roles to play in political movements—indeed, contemporary political movements are largely impossible without them (which is why television and radio stations are often the first targets whenever a revolutionary movement attempts to seize power)—there has never been a revolution conducted solely in cyberspace. Revolutions entail a taking to the streets and a taking of public space. They require the creation of disorder in places formerly marked by order and control. Political movements must take space and create it anew as a space in which the participants *can* be represented. While Lefebvre (1991) may theorize the continual production and representation of space and representational spaces, social movements understand that they must create spaces *for* representation (see D'Arcus 2001). The IWW knew this well (see Chapter 2), and so has every other important social movement, whether progressive or reactionary. The fascist movements (and governments) of 1930s Italy and Germany were adept at both taking space and transforming it into a new representational arena. While the taking and production of public space is a necessary component of democracy, it is not *only* or even *necessarily* democratic. Public space always entails risks; public space, like the disorder that must be part of it, is an inherently dangerous thing.[34] This is why, as we saw in Chapter 2, the protests outside of abortion clinics—and the means by which they are regulated—are so vexing.

Opponents of public, unmediated, and thoroughly politicized spaces, and of the disorder that must be part of them, have responded by creating a new "enclosure" movement every bit as undemocratic as that advocated by Matthew Arnold. Fearful of disorder and potential violence in public space, many developers, planners, and city officials (and planners of economic summits) advocate taming space by circumscribing the activities—and people—permissible within it. Powerful processes of exclusion are thus arrayed against the play of assertive, uncontrolled differences so necessary to democratic public spaces. As Lefebvre (1991, 373) has argued, differences threaten social order and hence must be absorbed by hegemonic powers:

> Differences arise on the margins of the homogenized realm, either in the form of resistances or in the form of externalities. . . . What is different is, to begin with, what is *excluded*: the edges of the city, shanty towns, the spaces of forbidden games, of guerrilla war, of war. Sooner or later, however, the existing center and the forces of homogenization must seek to absorb all such differences, and they will succeed if these retain a defensive posture and no counterattack is mounted from their side. In the latter event, centrality and normality will be tested to the limits of their power to integrate, to recuperate, or to destroy whatever has transgressed.

Whether challenged from the left or the right, the established power of the state and capital are threatened by the exercise of public rights within public space. The conflicting desires for order and for rights and representation—the need to go again to Hyde Park—structured the 1991 riots at People's Park. Activists in Berkeley fought on behalf of the expansion of social rights and opposition to outside control over the park: the power of the state and corporate capitalism, they felt, had to be opposed by (re)taking space. *Only* by taking and maintaining control over People's Park could oppositional political activity be represented and advanced. For activists such as David Nadle, the precedent was clear. The struggle in People's Park was another "Tiananmen Square" in which park activists and homeless people together would halt the expansion of the corporate state.

CONCLUSION: THE END OF PEOPLE'S PARK AS A PUBLIC SPACE?

The university seemed just as clear in its use of precedents. According to an unnamed university employee, Berkeley Chancellor Cheng-Lin Tien "personally rejected" the possibility of further negotiations with activists during the riots "on the grounds that he wanted violence and confrontation to show the regents he is tough. He alluded to Bush's actions in the Persian Gulf; you don't negotiate, you simply attack" (Kahn 1991c, 13). Attack was necessary because the occupation of People's Park by homeless people and activists was illegal and illegitimate and because that occupation had excluded the majority from the park. Berkeley City Manager Michael Brown promised that the city would do all that was necessary to ensure implementation of a more orderly vision of public space in People's Park. Referring to the homeless residents and activists, Brown told the *New York Times* (1991b, A8): "If they obstruct the majority opinion in a democracy, the city, the university, the county, and the state will apply whatever force is necessary to carry out the law." Brown kept his word. In the midst of the battle between protesters and police, Brown told the press: "We have a serious situation out there. People think this is about volleyball at the park but it is not. It's about a group of people who think they can use violence to force their will on a community, and we won't accept that" (Lynch 1991a, A21). "We almost lost the city," he added later (Kahn 1991c, 13); the police and the governing institutions of the city, according to Brown, were nearly incapable of quieting the disorderly politics of the street (Kahn 1991c).

The long-simmering and sometimes white-hot controversies over People's Park in Berkeley are paradigmatic of the struggles that define the nature of "the public" and public space. Activists see places such as the park as places for representation. By *taking* public space, social movements represent themselves to larger audiences. Conversely, representatives of mainstream institutions argue that public spaces must be orderly and safe in order to function properly. These fundamentally opposing visions of public space clashed in the riots over People's Park in

August 1991, and it is through such clashes that the actual nature of the right to the city is determined.

Though its "public" status remains ambiguous to this day (given UC's legal title to the land), the political importance of the park as a public space rests on its status as a *taken* space. By wresting control of the park from the state, park activists, to one degree or another, and over a period of more than 30 years, have held at bay those who wish to impose on the land a very different conceptualization of public space. But for those opposed to the park's continuing as some sort of "untamed land" (as the *Contra Costa Times* put it), the park's long-standing use as a refuge for homeless people suggested that it had become unmanageable, that large segments of the public felt threatened by the park's resident population, and that the city and the university needed to exercise more control over the park.

The riots that have occurred in and over the park—in 1969, in 1989, and again in 1991—require us to focus attention on exactly those issues Matthew Arnold so long ago pointed to: appropriate uses of public space, the definitions of legitimate publics, and the nature of democratic discourse and political action. Struggles over public space are struggles over opposing ideologies, certainly; but they are also struggles over the *practice* of democracy, a practice that is as often determined in the streets, on the sidewalks, and in the parks as it is in the halls of the legislature or in the courtroom. Oppositional movements, as well as movements seeking to create a new kind of space and a new kind of world, such as that which constructed People's Park in the first place, continually strive to assure the currency of more expansive visions of public space. Still, as we will see even more clearly in the next chapter, to the degree that the "disneyfication" of public space advances and both marginalized people and political movements are shut out of public space, the possibility of finding spaces that can be taken and made into a space for representing the right to the city seems to become ever more remote. That is why, as the activists that Naomi Klein (1999) profiles make clear, it is necessary to oppose the usurpation of public space and its privatization at every turn.

CODA

More than ten years after the 1991 riots, the fate of People's Park still remains unclear. The volleyball courts were built but were rarely used.

Residents of the park occasionally sabotaged them by burying broken glass in the sand. Eventually, under pressure from park users and the city, the university removed them altogether (Figure 4.6). The Free Stage and the Free Box are still there, despite frequent threats by the university and sometimes the city to remove them, and the basketball court is well used. The toilet facilities, which also house an equipment shed that has doubled as a police substation, were quickly covered with murals and graffiti (Figures 4.7 and 4.8). Numerous homeless people still sleep in the park, mostly at the east end under the trees. The west end contains several community garden allotments. The big grassy field is still there, and during the day groups of homeless people, mostly men, lounge about. Food Not Bombs provides regular meals (a practice opposed by many merchants and some of the neighborhood associations in the city).

Showing its typical bad sense of timing, in April 1999, as the 30th anniversary of the 1969 riots approached, the UC administration, in the person of Berkeley Chancellor Robert Berdahl, reiterated once again its desire to build student housing on the site of People's Park, declaring the park to be "underutilized and unsafe" (Burress 1999; Wong 1999). The president of the Berkeley Council of Neighborhood Associations supported Berdahl, saying that "the days of People's Park's historical significance are long gone" (Wong 1999, A22). Even so, she averred, the university should set aside a little space to commemorate what it once was (Burress 1999). Students, however, did not support the building of housing on the site. A year later, in a nonbinding resolution students voted 54% in favor of keeping the park as it is and not building dormitories (Lee 2000, A13). The vote came in the wake of the announcement of a new university long-range plan that held as one of its options the building of an "urban village" on the People's Park site.

When the fiscal year ended on June 30, 2000, the university sent a letter to the city announcing that it would no longer pay the nearly $200,000 a year in maintenance costs for the park as it had for the preceding 10 years under the 1991 agreement. The university would no longer clean and maintain the toilets or support the recreation programs the city developed for the park (Holtz 2000a). The university asserted that it could no longer afford the upkeep costs, but some city officials, such as one of the Parks and Recreation commissioners, felt that the withdrawal of funds had more to do with the city government's failure to be "tough enough on the poor" and homeless (quoted in Holtz 2000a, A15). In response to the university's decision and other con-

FIGURE 4.6. Part of the ongoing protests at the volleyball courts, 1994. The courts were eventually removed, but the basketball courts across the park have been retained. Photograph by Nora Mitchell; used by permission.

cerns, a number of park activists and supporters revived a plan to raise money to purchase the park and establish it as a land trust that would be administered by the city. Other activists countered that the university should continue its maintenance agreement while leaving the park as it is. As one of them put it: "I think it has been bought and paid for in blood" (quoted in Holtz 2000b, A18).

Whatever the fate of the park, it remains, for now at least, an intensely conflicted spot, and it is this very history of conflict that is important to how we understand—and act on—the relationship between public space and democracy, and how we determine just who counts as part of "the public." For, despite 30 years of confrontation, and despite Berkeley's ongoing gentrification and the continued development of the

FIGURE 4.7. The front of the restroom and utility building constructed in 1991. It was almost instantly covered with murals and graffiti, much of it depicting the various riots at People's Park and ongoing concerns about police brutality in the city. Photograph by author.

FIGURE 4.8. The rear of the restroom and utility building. The area behind the rolling door was used for sports equipment storage and checkout and as an occasional police substation. Photograph by author.

Berkeley campus and the surrounding neighborhoods, People's Park remains, troubled as it is, a refuge for homeless people, people who have no other place just to be. As the next chapter will show, such places are increasingly rare, and their destruction has clear implications for just who has the right to the city.

NOTES

1. The defense of the stage and the Free Box are in fact quite important and indicate why People's Park is such an important space in the current history of American public space. The stage was built explicitly as a space for free speech and political action, and it has remained a key center for rallies and organizing efforts in the city. In this sense, People's Park was constructed as a public space for politics, as a place where political involvement and debate were encouraged—and in a way that stood at odds with (but not disconnected from) the more orderly politics of the traditional parties, elections, council meetings, and the like. The "Free Box" is a clothes (and other materials) exchange. People leave what they no longer need for others to pick and choose as they please. The Free Box is a fully decommodified system of exchange of use values (to put it in technical terms), and as such represents the possibility of public space as a noncommodified space in the city where people can meet their needs in a manner not entirely predicated on capitalist relations of property, exploitation, and exchange value. Whatever the differences between the politics of protest and the politics of homelessness in the American city, they are united in their need for a *public* space either relatively free or liberated from the controlling power of the state and property. This chapter will begin to show just how complex the "relatively" in the previous sentence is.
2. In 1996 Nadle was murdered in his nightclub.
3. With these comments, Nadle makes it clear how the struggle for People's Park foreshadowed some of the key issues that were to become radical battlegrounds later in the 1990s and into the next decade, including the corporate dominance of public space (see Klein 1999). Organizations such as Reclaim the Streets, Critical Mass, and the coalitions that have disrupted world trade meetings have expressed a deep affinity for the Berkeley activists of the 1980s.
4. My evidence here comes from my brother, David Mitchell, who was one of those neighboring students and a supporter of park redevelopment. His constant questioning of my positions in my research on People's Park has been invaluable to its development.
5. The best reporting on the riots is in the weekly *East Bay Express* (Auchard 1991, 1ff; Kahn 1991c, 1ff; Rivlin 1991b, 1ff), which details incidents of police abuse and the actions of protesters.

6. I recognize that there are potentially many more ways of seeing public space (some of which will be explored in later chapters) and that many people will hold a middle (and perhaps wavering) ground between them. But these, as we will see, are the predominant ways of seeing public space across a variety of (largely Western) societies and historical periods. I suggest in what follows that in examining these visions we can begin to see how public space is produced through practices guided or structured through their dialectical interaction.

7. Lefebvre (1991, 39) claims that representational space is "passively experienced" by its users. This thesis will not withstand close scrutiny. People actively transform their spaces, appropriating them (or not) strategically.

8. Some critics of my position here (cf. Heyman 2001) argue that *taking* space is fundamentally different from *producing* it. Heyman's argument is that any public space must be a *new kind* of space that represents new social relations, not a space transformed. At the level of a philosophical thesis, Heyman is perhaps correct. In the physical world, however, it is hard to see how physical space can be conjured up out of nothingness: space already exists, and indeed *must* exist, in order for it to be socially produced as *public* space.

9. That this is the case is now very well understood by planners of major international summits (such as WTO meetings). It is now standard practice to do all that can be done to render protests against international ministerial meetings invisible by locating the meetings behind miles-long fences or in the heavily guarded compounds of totalitarian states (such as Qatar). This is the face of democracy under globalization.

10. This is not to say that public space is sufficient—only that it is necessary. Sexual minorities, for example, often have very real needs for *private* space—space free from the surveilling eyes of the state or dominant society—in order to both fulfill desires and to fashion identities. Yet, it is also the case that such minorities have become *political* actors to the degree they have forced themselves into public space, as with ACT-UP, Queer Nation, or through the development of spatially concentrated neighborhoods where gay men and women and other sexual dissidents are regularly seen in public on the streets. For recent discussions of these issues see Brown (2000) and Hubbard (2001).

11. As we will see in the next two chapters, understanding the rights of the homeless to be *seen* in cities in this manner sheds important light on some of the consequences (if not always the intent) of anti-homeless laws. Anti-homeless laws have the effect not only of regulating homeless people's behavior but also of delegitimizing them as bonafide members of the public.

12. I will save a fuller discussion of what constitutes "danger" for Chapter 5, in which we will examine the ideal of what Sennett (1994) calls an urban environment free from "resistance."

13. Young (1990, 119) goes on to argue that, in order to promote a democratic politics of inclusion, "participatory democracy must promote the ideal of a

heterogenous public"—exactly that, as I will later argue, which so much current public space law and planning does *not* promote.

14. At least that was how the ideal of separate spheres posited the relationship between them. The degree to which this ideal was matched in practice can be gauged quite precisely by reading the 1848 Declaration of Sentiments that launched the feminist movement in the United States—a document that spells out how women were excluded from the polity (and turned into privately held property) while at the same time showing how some women were able to break the bounds of that ownership and exclusion.

15. Even if their labor, and often their bodies, have been quite welcome. Women in public as other than decorations on the arms of men (or working-class cleaners, sellers, waitresses, etc.), as E. Wilson (1991) has shown, have historically been viewed as suspicious, as prostitutes, or as deranged and uncontrollable. Alternatively, stylized representations of women in public—the bare-breasted heroine on the barricades—have played an ideologically important role in struggles over public space.

16. The argument here is similar to Benedict Anderson's (1991) famous argument about the nation being an "imagined community," since the community in total can never be experienced or even known.

17. Numerous geographers have attempted to advance remarkably unmaterial definitions of public space, but they have yet to show how such definitions differ in any substantial way from the rather ethereal sense that attaches to the public sphere. As importantly, losing focus of the materiality of public space leads to a rather remarkable inability to engage in clear empirical analysis of the spatiality of political events. See Kilian (1998) and Ruddick (1996).

18. Howell (1993) notes that the difference between Arendt and Habermas is that, for the former, public space has not lost its "geographical significance."

19. For a fuller analysis of the relationship between public and private *activities* and public and private *spaces*, see Staeheli (1996).

20. The ideological and definitional exclusion of homeless people from the polity has a long history in Anglo-American jurisprudence (see Ribton-Turner 1887; Commonwealth of Pennsylvania 1890). It finds its practical expression in such mundane things as laws that require a fixed address in order to register to vote.

21. In such discourse "the homeless" are presented as a homogeneous mass, with few if any characteristics to distinguish them—in essence denying homeless people's individuality and humanity. This discourse operates simultaneously with another that seeks to particularize the homeless, showing that this one is an alcoholic, that one a drug addict, still another is mentally ill, and a fourth is all three. The strategy here is to so particularize the homeless as to deny what is common among them: namely, that they are without permanent shelter of their own. I take on this latter discourse (to some extent) in the next two chapters; but throughout this volume I also engage in my own essentialization of "the homeless." This is a purely

political choice on my part: it is a means to focus on political processes and struggles that shape and define the homeless as a class with a set of common interests rather than as pathological individuals needing treatment or other forms of paternalistic intervention.

22. Goss (1996, 1999) has softened his stance a bit, focusing now as much on the ways that users of highly designed spaces transform those spaces while they are using them often into things their designers did not plan. Lees (1998, 2001) argues that some new urban public spaces—such as the Vancouver Public Library—are in fact fairly political places despite, or even because of, their security apparatus. And studies of consumers using malls (Miller, Jackson, Thrift, Holbrook, & Rowlands 1998) argue that consumers engage in tactical appropriation of meanings while they shop. All of these studies argue that meanings of particular landscapes are never fixed once and for all, which is true, but they tend to gloss over the degree to which different actors possess different levels of power to determine both uses and meanings of public space, a point very clearly made in the introduction to Gold and Revill (2000).

23. This seems especially apparent in new civic buildings, such as Vancouver's new library, according to the evidence, if not the analysis, presented by Lees (2001).

24. This point is also central to Debord's (1994 [1967]) theory of spectacle. Debord's argument undergirds much of what follows.

25. Compare Wallace (1996), who argues that the presentation of spectacle in place of history and society fits well with prevailing corporate conceptions of progress and "democracy."

26. So pervasive has this suburban sensibility become that architects and designers seem to find it nearly impossible to transcend even when facing the prospect of designing a new *urban* park such as Toronto's Downsview (Mitchell and Van Deusen 2002).

27. Cindi Katz's (1998) examination of Central Park in this regard is especially important.

28. This was another trend amplified in the wake of the September 11, 2001, terrorist attacks. In the massive reporting on the event, not a little ink and airtime was devoted to what was being said—what political arguments were being made, what rumors were flying, etc.—on the Internet.

29. MMCh analysis and the MTV campaign are two examples of the massive outpouring of faux populism issuing from corporate boardrooms, ad agencies, and cultural studies centers that Thomas Frank (2001) so thoroughly demolishes in *One Market, Under God*.

30. See Rosati (2002) for an excellent analysis of these practices.

31. Sennett (1992) provides an excellent analysis of the dangers of this sort of solipsistic empowerment.

32. It is also, as we saw (on pp. 71–72), hemmed in by some quite bad constitutional precedent that makes it a particularly anemic "public forum."

33. Hershkovitz is arguing, rightly, against de Certeau's (1984, xix) notion that popular movements are necessarily "placeless"—that hegemonic powers

have monopoly power over space and place, and hence resistance can occur only in the interstices—that is, it can *only* be "placeless."

34. In the next chapter I introduce the idea of a "brutal public sphere." The
dangers of public space should not be equated with this brutal public
sphere. "Public space," in this case, refers to an environment of risk, the
risk necessary to any democratic politics. But "public sphere" refers to sys-
tematic oppression and exploitation, either organized by the state or by
private interests.

5

The Annihilation
of Space by Law
Anti-Homeless Laws and the Shrinking
Landscape of Rights

No one is free to perform an action unless there is somewhere
he is free to perform it. . . . One of the function of property
rules, particularly as far as land is concerned, is to provide a
basis for determining who is allowed to be where.
—JEREMY WALDRON (1991, 226)

When some members of the Berkeley City Council feared that the university had ended its decade of support for People's Park because the
City had not been "tough enough" on the homeless and other poor people, one must wonder just what constitutes "toughness" these days. For
Berkeley, though often experimenting with "liberal" policies toward the
homeless (such as establishing a program in which pedestrians gave
vouchers for services to panhandlers in lieu of money), has been one of
the leaders of a new legal assault on homeless people. This assault takes
the form of passing and implementing a suite of "quality of life" initiatives and laws that seek to highly regulate street behavior, when and
where (or if) people can sleep in public, and how people can and cannot
beg. In 1994, the Berkeley City Council and voters approved an anti-
panhandling law that prohibited "aggressive" panhandling, all begging
at night, panhandling people as they got in and out of their cars, and
begging inside a 10-foot "bubble" around every automatic teller ma-

chine. In addition, the law made it illegal to sit on the sidewalk. Berkeley's regulations were "among the strictest in the country" (*San Francisco Chronicle* 1994a). When the Berkeley regulations were held up in the courts for several years (Herscher 1995), a new, slightly, revised version of the laws was passed by the city council in 1998.

The Berkeley ordinance is part of a species of law and policy that has developed over the past decade in response to the permanent crisis of homelessness. A few examples suffice to give a sense of what is at work:

- Over 11 months in 1998, San Francisco issued more than 16,000 "quality of life" violations, mostly to homeless people (NLCHP 1999). As part of its "Matrix" program, the city frequently "sweeps" city streets, squares, and parks of homeless people, and enforces a "zero tolerance" policy for violations of laws against camping in public, loitering, urinating and defecating in public, and drinking in public places (MacDonald 1995).
- In dozens of cities around the country, including Santa Cruz, Berkeley, Phoenix, St. Petersburg, and San Diego, it is illegal to sleep in public places. Similar laws are on the books in Seattle where they were used to arrest protesters at the December 1999 WTO meeting (Falit-Baiamonte 2000).[1]
- In Atlanta and Jacksonville, Florida, it is a crime to cut across or loiter in a parking lot (in 1 month alone in 1993 in Atlanta, 226 people were arrested for "begging, criminal trespass, being disorderly while under the influence of alcohol, blocking a public way or loitering in a parking lot" [*Atlanta Journal and Constitution* 1993]).
- In New York it is illegal to sleep in or near subways and to wash car windows on the city streets (Howland, 1994); most of the small squares in Manhattan have now been leased to Business Improvement Districts, whose private security forces vigorously enforce rules against dozing on benches or at tables (cf. Katz 2001).
- In Eugene, Oregon, and Memphis, Tennessee, beggars are required to obtain licenses, a process that requires being fingerprinted and photographed. Beggars are required to carry their photo-licenses at all times (*San Francisco Chronicle* 1994b).
- As in Berkeley, in Cincinnati it is illegal to beg from anyone getting into or out of a car, near automatic teller machines, after 8 P.M., or within 6 feet of any storefront. It is also illegal to sit or

lie on sidewalks between 7 A.M. and 9 P.M. (*Cincinnati Enquirer* 1995a, 1995c).[2] More than a third of all municipalities in the United States now have such laws (*Denver Post* 1999a). Denver, Colorado, enacted its law, in the words of the president of its Downtown Partnership, because "Panhandling makes [visitors] cringe, especially if they don't know where they are" (*Denver Post* 2000a).

- Not content with these sorts of restrictions on the place and manner of begging, Baltimore is seeking to ban panhandling altogether after dark. Advocates of the new law there say that nighttime panhandlers disrupt people who want "to go to Little Italy at night to dine or to Fells Point to barhop" (*Baltimore Sun* 2001).

The intent is clear: to control behavior and space such that homeless people cannot do what they must do in order to survive without breaking laws. Survival itself is criminalized. But, as legal scholar David Smith (1994a, 495) argues in an article on the criminalization of homelessness, the "supposed public interest that criminalization is purported to serve"—such as the prevention of crime or the maintenance of order—"is dubious at best," since criminalizing necessary behaviors does nothing to address such root causes as the lack of affordable, safe housing in most cities, structural unemployment (or, as I would put it, the need to maintain a reserve army of the unemployed), and the pairing of poverty and despair that turns drug and alcohol addiction and mental illness into an issue of *housing* for a significant portion of the population. If Smith is correct, then two questions arise. First, just why have such anti-homeless laws become so prevalent in the past decade (and why are they continually touted as the key to "saving" America's cities)? Second, what do such laws portend for urban public spaces, and the practices of democracy and citizenship that such spaces do and do not allow? This chapter seeks to answer those questions. In the process of doing so, it shows that the "cry and demand" for the right to the city must become ever more insistent.

THE ANNIHILATING ECONOMY

For Neil Smith (1996), the rise of anti-homeless laws, coupled with a range of other punitive laws including so-called welfare reform, announces a new urban regime. This regime, Smith argues, is based on "revanchism," a right-wing movement of "revenge" for the presumed "excesses" of the liberal 1960s that seeks to revive what it sees as the "traditional values" of America.[3] But, as Smith is quick to point out, this revanchism is not solely a right-wing movement. Indeed, some of its most infamous moments, such as the closing of Tompkins Square Park, were the result of liberal urban administrations (N. Smith 1996, 220). Indeed, what is at work is the implementation, at the urban scale, of a regulatory regime—and its ideological justification—appropriate to the globalizing neoliberal political economy that developed out of the global recessions of the 1970s, the debt crisis of the early 1980s, the economic crises of the late 1980s (and 1990s, for Asia), and the implosion of the Soviet Union and its satellites. Such liberal mayors as Paul Schell in Seattle, David Dinkins in New York, or Willie Brown in San Francisco have been no less insistent on the need to reregulate the poor and homeless than their conservative and even reactionary brethren, such as Dinkins's successor in New York, Rudy Giuliani, Brown's predecessor in San Francisco, Frank Jordan, or Schell's once-hopeful successor in Seattle, City Attorney Mark Sidran (whom we will get to know a little bit better in a moment). "Revanchism" describes an urban regime that cuts across mainstream party lines and has even taken on the cast of common sense. It is a powerful movement reacting to what seems to be a powerful set of trends shaping urban areas, trends that are organized under the capacious banner of "globalization."

On the one hand, globalization refers to the process of integration of economies across international boundaries. On the other hand, it refers to the sense that, not just in economic terms, but also in social, political, and cultural terms, the world is ever more connected—that time and space are constantly being compressed (Harvey 1989, 1990; Massey 1995) and places and borders are of decreasing importance. In this respect, globalization is, more than anything, a remarkably powerful ideology in and of itself. Indeed, the popular media remain enthralled by the prospect of "globalization" (despite the growing global protest movement against the form it is taking, and despite the "blowback" that the September 11, 2001, terrorist attacks in part represented), breathlessly recounting the wonders it is leading to: instant communication, Big Macs available in every corner of the world, the sublime joy of being able to eat sushi in the middle of a Nebraska winter. Watchers of the news and readers of the papers are led to believe, simply, that space has ceased to exist (despite occasional troublesome hiccups, such as a dozen undocumented migrants dying in the Arizona desert as they seek to elude the border patrol).

For the class the ideology of globalization serves, "globalization" is little more than an accurate moniker for the new experience of everyday life: the managerial elite who play and govern instantaneous markets in currency, futures, stocks, and even inventories; the Western and westernized middle and upper classes that can afford both the equipment and the time to instantly connect to the far corners of the globe through the World Wide Web; the wealthy students who jet across continents for long weekends with relatives or friends or skiing holidays in the fresh powder of the Rockies or Alps. For such people, space simply does not matter (at least once they have recovered from jet lag). Or better, it simply is not matter: it is rather some ethereal medium made increasingly irrelevant by networks of wires, fiber-optic cables, superhighways (asphalt or informational), jetliners, satellites, and, of course, money. A large number of people and, more importantly, capital itself have been unfettered (and perhaps just a bit disoriented) by time–space compression. There seems to have been, to use one of David Harvey's favorite insights from Marx, a further, and quite incredible, "annihilation of space by time." Even those who adopt a critical stance toward globalization (such as Harvey himself) but nonetheless see the "annihilation of space by time" as the overriding economic force of our era still tend to see capital as a global, translocal force able to behave, in Smith's (1990) imagery, like a plague of locusts circling the globe, touching down hither and yon, devouring whole places as it seeks ever better comparative advantage.

Yet, as a number of geographers have shown, such a globalization is in fact *not* predicated on the "annihilation of space by time," no matter how evocative that metaphor may be, but rather on the constant production and reproduction of certain *kinds* of spaces (Harvey 1982; Massey 1995; Storper and Walker 1989; N. Smith 1990, 1996; Walker 1996). For capital to be free, it must also be fixed in place. This is the central geographical contradiction of capitalism, and the one that makes the *ideology* of globalization, together with neoliberal, revanchist social regulation, so important. It is a contradiction that is rooted in capitalism's tendency toward a continually declining rate of profit. "Going global"—reconfiguring the spaces and scales of accumulation—is one means of staving off that decline, at least for some capitals, some of the time. Not just at the global scale, but in all the locations that capital does business, perpetual attempts to stave of crisis by speeding up the circulation of capital leads to a constant reconfiguration of productive relations (and productive spaces). Together, these trends—toward rapid turnover and toward the concomitant appearance of globalization—create a great deal of instability for those whose investments lie in fixed capital, especially the fixed capital of the built environment.[4] While capital simply cannot exist without some sort of fixity—in machines and factories, in roads and parks, in homes and stores—the very unevenness of capital mobility lends to places an increasing degree of uncertainty. Investment in property can be rapidly devalued, and local investors, property owners, and tax collectors can be left holding the bag. Or not. Together or individually, they can seek to stabilize their relationship with peripatetic capital by protecting long-term investments (and attracting new investment) in fixed capital—and as a home-base for globe-trotting capital—through tax, labor, environmental, and regulatory inducements. But establishing incentives and transforming regulatory environments can lead to a frenetic place auction, as new municipalities and states compete with on another both to attract new investment and to keep local capital "home."

This is precisely where the ideology of globalization is so powerful: by effectively masking the degree to which capital must be located, the ideology of globalization allows local officials, along with local business and property owners, to argue that they have no choice but to prostrate themselves before the god Capital, offering not just tax and regulatory inducements but also extravagant convention centers, downtown tourist amusements, up-market, gentrified restaurant and bar districts, new baseball and football stadiums, and even occasional investment in such amenities as museums, theaters, and concert halls (Molotch 1976; Cox and Mair 1988, Zukin 1995).[5] Image becomes everything. When capital is seen to have no need for any particular place, then cities do what they can to make themselves so attractive that capital—in the form of new businesses, more tourists, or a greater percentage of suburban spending—will *want* to locate there. If there has been a collapse of space, then there has also simultaneously been a new and important reinvestment in *place*—a reinvestment both of fixed (and often collective) capital and of imagery. For Scott Kirsch (1995, 529) a world thus structured leads to the obvious question "What happens to space *after* its collapse; how do these spatiotemporal transformations impact our everyday lives . . . ?"

For many cities in the United States, the answer to this question, quite perversely, has led to a *further* "annihilation of space"—this time not at the scale of the globe and driven by technological change, but now quite locally and driven by changes in law. New laws governing the use of space are not just a rhetoric or discourse of neoliberal revanchism, but its actual practice, a practice that is a key front upon which the battle for the right to the city must be fought. In city after city concerned with "quality of life"—with, in other words, making urban areas attractive to both footloose capital and to the footloose middle and upper classes—politicians and managers of the new economy have turned to what could be called "the annihilation of space by law"—the space to live, sit, and take care of oneself if there is no house or home in which to do so. For this is what the new legal regime in American cities—the regime that is represented in the sorts of laws described above—is outlawing: just those behaviors that poor people, and the homeless in particular, must do in the public spaces of the city.[6] And this regime does it by legally (if in some ways figuratively) annihilating the only spaces the homeless have left. The anti-homeless laws being passed in city after city in the United States work in a pernicious way: by redefining what is acceptable behavior in public space, by in effect annihilating the spaces in which homeless people *must* live, these laws seek simply to annihilate homeless people themselves, all in the name of re-creating the city as a playground for a seemingly global capital that is ever forced to engage in its own annihilation of space.

THE ANNIHILATION OF PEOPLE BY LAW

Sleepless in Seattle

The current restrictions on homeless people's behavior in public space are clearly an effort to regulate space so as to eliminate homeless people, not homelessness. Berkeley is quite advanced in this effort, though not nearly so much as San Francisco across the bay or Seattle to the north. The case of Seattle, in fact, is indicative of the whole tenor of the war against homeless people that cities are waging in the name of global competitiveness. It is also indicative of the tortured path that ideology travels as it is transmogrified from a form of urban liberalism into a form of urban neoliberalism.

In an article that recounted the Disney Corporation's failed attempts at urban planning in Seattle during the 1980s and the subsequent development by the city of a more inclusive planning process— one that included homeless people and attempted to incorporate their needs and desires—Stacy Warren (1994) quotes the remarks of a homeless man included in a 1989 survey: "thank you for having me and other individuals to be part of the [Seattle] Center–warmth, etc. as a homeless person." On the basis of this and other evidence, Warren (1994, 110) concludes: "That a homeless person, as perhaps the strongest symbol of disenfranchisement in the city, should form a constituent part of the planning process for the new Seattle Center speaks to the power of true citizenship embedded in hegemonic processes."

Such benevolence toward homeless people in Seattle had its limits, even in the 1980s. In 1986 Seattle passed an "aggressive panhandling law" (*Los Angeles Times* 1987; *New York Times* 1987; Blau 1992), one of the earliest in the country. The law was struck down as unconstitutional (since is was seen as outlawing a form of protected speech). By the early 1990s, unconvinced of the effectiveness of laws merely regulating aggressive panhandling, Seattle's crusading City Attorney Mark Sidran sponsored a suite of new laws that outlawed everything from urinating in public to sitting on sidewalks and sleeping in public places. The new laws further gave the police the right to close to the public any alley it felt constituted a menace to public safety.[7] Sidran argued that such laws (that is, laws that outlawed conduct that homeless people had to engage in to survive) were necessary to assure that Seattle did not join the cities of California as "formerly great places to live." The danger was palpable if still subtle:

> Obviously, the serious crimes of violence, the gangs and the drug trafficking can tear a community apart, but we must not underestimate the damage that can be done by a slower, less-dramatic but nonetheless dangerous unraveling of the social order. Even for hardy urban dwellers, there comes a point where the usually tolerable "minor" misbehaviors—the graffiti, the litter and the stench of urine in doorways, the public drinking, the aggressive panhandling, the lying down on the sidewalks—cumulatively become intolerable. Collectively and in the context of more serious crime, they create a psychology of fear that can and has killed other formerly great cities because people do not want to shop, work, play or live in such environments. (Sidran 1993, B5)

The logic is fascinating. It is not so much that "minor misbehaviors" are in themselves a problem. Rather, the context within which these behaviors occur ("more serious crime") makes them a problem. Sidran is expressing a variation on the "broken windows" thesis of James Q. Wilson and George F. Kelling (1982; Kelling and Coles 1996), which we will explore in more detail in the next chapter. For now it is enough to note that the answer to the problems associated with "minor misbehaviors" and their context of "serious crime" is neither to focus on the context nor to try to understand the reasons *why* people might need to lie on sidewalks or urinate on doorsteps. Instead, "[t]o address misbehavior on our streets, we need to strengthen our laws. We need to make it a crime to repeatedly drink or urinate in public, because some people ignore the current law with impunity" (Sidran 1993). Sidran recognizes that "law enforcement alone is not the answer" and thus supports expanded services for the homeless. "At the same time, however, more services alone are also not the answer. Some people make bad choices"—such as the choice to urinate in public or to sit on sidewalks. "We also need to address those lying down day after day in front of some of our shops. This behavior threatens public safety. The elderly, infirm and vision impaired should not have to navigate around people lying prone on frequently congested sidewalks."

There is another, perhaps more important, danger posed by those sitting and lying on sidewalks: "many people see those sitting or lying on the sidewalk and—either because they expect to be solicited or otherwise feel apprehensive—avoid the area. This deters them from shopping at adjacent businesses, contributing to the failure of some and damaging others, costing Seattle jobs and essential tax revenue" (Sidran 1993). Sidran argues in the end that homeless people in the streets and parks "threaten public safety in a less-direct but perhaps more serious way. A critical factor in maintaining safe streets is keeping them vibrant and active in order to attract people and create a sense of security and confidence." And security is precisely the issue:

> If you were to write Seattle's story today, you might borrow Dickens's memorable opening of "A Tale of Two Cities," "It was the best of times, it was the worst of times." From *Fortune* Magazine's No. 1 place to do business to the capital of "grunge," from high-tech productivity perched on the Pacific Rim to espresso barristas on the corners, it is the best of times in Seattle. We're even a good place to be sleepless.[8]

Especially if you are homeless, it seems, since under Sidran's proposal there would simply be no place for you to sleep. The regulation of public space takes on a different "caste," however, depending on who you are. Under Sidran's proposals (eventually passed by the Seattle City Council), exceptions to the "no sitting" provisions were made for "people using sidewalks for medical emergencies, rallies, parades, waiting for buses or sitting at cafes or espresso carts" (*Seattle Times* 1993a). The target of these laws is obvious.[9] And their effect was both predictable— when enforcement was emphasized downtown, many homeless people moved to outlying business districts, prompting numerous complaints from merchants in those areas (Balter 1994)—and important to understand. To the degree that laws can annihilate spaces for the homeless (the sidewalk, the park, the alley), they can annihilate the homeless themselves. When such anti-homeless laws have come to cover all public space, which is certainly the hope of residents and merchants in outlying areas when downtown ordinances push homeless people in their direction, then presumably the homeless will just vanish.

The Annihilation of People

This is the crux of the matter. Arguing from first principles in a brilliant essay, the legal scholar Jeremy Waldron shows that the condition of being homeless in capitalist societies is most simply the condition of having no place to call one's own. "One way of describing the plight of the homeless individual might be to say that there is no place governed by a private property rule where he is allowed to be" (Waldron 1991, 299). Homeless people can only be on private property—in someone's house, in a restaurant's toilet—by the express permission of the owner of that property. While that is also true for the rest of us, the rest of us nonetheless have at least one place in which we are (largely) sovereign. We do not need to ask permission to use the toilet or shower or to sleep in a bed. Conversely, the only place homeless people may have even the possibility of sovereignty in their own actions is on common or public property.[10] As Waldron explains, in a "libertarian paradise" where *all* property is privately held, a homeless person simply could not *be*. "Our society saves the homeless from this catastrophe only by virtue of the fact that some of its territory is held as collective property and made available for common use. The homeless are allowed to *be*—provided they are on the streets, in the parks, or under bridges" (Waldron 1991, 300).

Yet, as city after city passes laws specifically outlawing common behaviors (urinating, defecating, standing around, sitting, sleeping) on public property:

> What is emerging—and it is not just a matter of fantasy—is a state of affairs in which a million or more citizens have no place to perform elementary human activities like urinating, washing, sleeping, cooking, eating, and standing around. Legislators voted for by people who own private places in which they can do these things are increasingly deciding to make public places available only for activities other than these primal human tasks. The streets and subways, they say, are for commuting from home to office. They are not for sleeping; sleeping is what one does at home. The parks are for recreations like walking and informal ball-games, things for which one's own yard is a little too confined. Parks are not for cooking or urinating; again these are things one does at home. Since the public and private are complementary, the activities performed in public are the complement of those performed in private. This complementarity works fine for those who have the benefit for both sorts of places. However it is disastrous for those who must live their whole lives on common land. If I am right about this, it is one of the most callous and tyrannical exercises of power in modern times by a (comparatively) rich and complacent majority against a minority of their less fortunate fellow human beings. (Waldron 1991, 301–302)[11]

In other words, we are creating a world in which a whole class of people cannot be—simply because they have no place to be.

As troubling as it may be to contemplate the necessity of creating "safe havens" for homeless people in cities (and we will see just how troubling that is in the next chapter),[12] it is even more troubling to contemplate a world without them. The sorts of actions we are outlawing— sitting on sidewalks, sleeping in parks, loitering on benches, asking for favors, peeing—are not themselves subject to total societal sanction. Indeed, they are all actions we regularly and even necessarily engage in. What is at issue is where these actions are done. For most of us, a prohibition against asking for a donation on a street corner is of no concern; we can sit in our studies and compose begging letters on behalf of the PTA or even ourselves. So too do rules against defecating in public seem entirely reasonable. When one of us—the housed—finds himself or herself unexpectedly in the grips of diarrhea, for example, the question is one of timing—not at all of having no place to take care of our needs. Not so for the homeless, of course: the homeless person with diarrhea is entirely at the mercy of property owners or must find a place on public property on which to relieve him- or herself. So too with the everyday need to defecate. And similarly, the pleasure (for me) of dozing in the sun on the grass of a public park is something I can, quite literally, live without, but only because I have a place where I can sleep whenever I choose. The issue is not murder or assault, in which there are (near) total societal bans. Rather, the issue, in the most fundamental sense, is an issue of geography, a geography in which a local prohibition (against sleeping in public, for example) becomes a total prohibition (for example, on sleeping) *for some people*. That is why Jeremy Waldron (1991) understands the promulgation of anti-homeless laws as fundamentally an issue of freedom: such laws destroy whatever freedom homeless people have, as people, not just to live under conditions at least partially of their own choosing, but to live at all.[13] And that is why what we understand public space to be, and how we regulate it, is so essential to the kind of society we make. The annihilation of space by law is unavoidably (if still only potentially) the annihilation of *people*.

The degree to which anti-homeless legislation diminishes the freedom or rights of homeless people is not, of course, an important concern for those who promote anti-homeless laws. Rather, they see themselves not as instigators of a pogrom but rather as saviors: saviors of cities, saviors of all the "ordinary people" who would like to use urban spaces but simply cannot when they are chockful of homeless people lying on sidewalks, sleeping in parks, and panhandling them every time they turn a corner. These are our latter-day "Little Arnolds," and theirs is not simply a good or just cause; it is a necessary one. "The conditions on our streets are increasingly intolerable and directly threaten the safety of all our citizens and the economic viability of our downtown and neighborhood districts," according to Seattle's Mark Sidran (*Seattle Times* 1993c). Or, as columnist Joni Balter (1994) put it: "Seattle's tough laws on panhandling, urinating and drinking in public, and sitting and lying on the sidewalk are cutting edge stuff. Anybody who doesn't believe in taking tough steps to make downtown more hospitable to shoppers and workers wins two free one-way tickets to Detroit or any other dead urban center of their choice." The argument couldn't be clearer. Urban decline is the result of homelessness. Detroit is "dead" because people "make bad choices" and panhandle on the streets, urinate in public, or sit on sidewalks, thereby presumably scaring off not only shoppers, workers, and residents, but capital too. This is a monumentally ignorant view of urban political economy (and, for that matter of racism in the United

States), but it is not at all an uncommon one. Without the elimination of homeless people, Seattle will go the way of Detroit and Newark; hence, the homeless must be eliminated.

THE PROBLEM OF REGULATION

While the mode of regulation of homeless people proposed by Seattle's Sidran and Berkeley's City Council may be relatively new, the desire to regulate the homeless out of existence is not. Indeed, what is at work in American cities is a *recriminalization* of homelessness. The criminalization of poverty and homelessness has a long history, of course. Aspects of the Elizabethan poor laws (which, as Marx showed, were so crucial to the rise of capitalism and the development of a reserve army of labor) were transferred to America and helped shape how colonial cities regulated the poor. During the depressions of the 1870s, 1890s, and early 1900, and their associated "tramp scares" (Cresswell 2001), American varieties of English poor laws were revived. And, as Piven and Cloward (1992) showed so well, welfare and other policies for "regulating the poor" have been an integral aspect of the American state's 20th-century desire to mediate the social pressures that arise from capitalism's fluctuating booms and busts. But all this regulating of the poor should not blind us to their absolute *necessity* to actually existing capitalism. In a striking passage, Marx (1987 [1867]) discusses the growth of the very poor as a function of the accumulation of capital. But he also points to the contradiction that this dual growth (of wealth and poverty) leads to:

> The greater the social wealth, the functioning of capital, the extent and energy of its growth, and, therefore, also the absolute mass of the proletariat and the productiveness of its labour, the greater is the industrial reserve army. The same causes which develop the expansive power of capital, develop also the labour-power at its disposal. The relative mass of the industrial reserve army increases therefore with the potential energy of wealth. But the greater this reserve army in proportion to the active labour-army, the greater is the mass of a consolidated surplus-population, whose misery is in inverse ratio to its torment of labour. The more extensive, finally, the lazarus-like layers of the working-class, and the industrial reserve army, the greater is official pauperism. *This is the absolute general law of capitalist accumulation.* Like all other laws it is modified in its working by many circumstances. . . . (p. 603)

Chief among these circumstances is the simple fact that "paupers" often simply will not stand for the status they are assigned, and this becomes a problem of social regulation, which may itself take on a particular historical logic. The very existence of such an army of poverty, which is so necessary to the expansion of capital, means there is an army of humanity that must be strictly controlled or else it will undermine the drive toward accumulation. If this has been a constant fact of capitalist development, then what sets the present era, and the present wave of anti-homeless laws, apart is the degree to which such regulation has also become an important ingredient in not just expanding capital but in either attracting it in the first place or in protecting it once it is fixed in particular places. This is what anti-homeless laws are meant to do. The contradiction, then, is that the homeless and poor are desperately *needed*, but not at all *wanted*, and so the solution becomes a geographical one: regulating space so that homeless people have no room to be *here*.[14]

In the mid-1980s, Andrew Mair (1986, 351) made a similar claim—about the necessity of removing homeless people from contemporary urban centers so as to assure their continued viability as sites for capital accumulation—for the case of Columbus, Ohio. He suggested that "while the removal of the poor may appear merely incidental with respect to urban redevelopment . . . it can be argued that the poor must *necessarily* be removed for post-industrial development to occur." But necessary as it may be, it is abundantly clear that as long as removal depended on the relocation of services (as described by Mair 1986; see also Dear and Wolch 1987; Takahashi 1998; Wolch and Dear 1993), it has not really worked. Closing down and relocating soup kitchens and shelters in city after city—or the creation of service-dependent ghettos (Dear and Wolch 1987) in marginal parts of the city—proved at best a temporary solution as more and more homeless people came to colonize the streets of downtown business and commercial districts. Excluded from housing by the destruction of single-room-occupancy hotels and other inexpensive housing (Baum and Burnes 1993, 139; Blau 1992, 75; Groth 1994; Hartman 1987; Hopper and Hamberg 1984; Kasinitz 1986); marooned by the retrenchment from federally subsidized housing for the poor beginning with the Carter administration, reaching full steam during the Reagan years, and fully consummated in the Clinton administration (Leonard, Dolbeare, and Lazere 1989; Crump 2002, 2003); made redundant by a quickly shifting economy that has seen real wages stagnate and even decline for most workers even during an economic boom; and thrown onto the streets through deinstitution-alization unaccompanied by a concomitant commitment to community-based care (Wolch 1980; Dear and Wolch 1987; Wolch and Dear 1993), homeless people turned to begging, hanging out, sleeping on the very streets they were meant to be excluded from. Similarly, these years saw fledgling movements by homeless people to protest their attempted ex-clusion from public space. When, in 1993, the Santa Monica City Coun-cil considered enforcing a law closing public parks from midnight until 5 A.M., for example, organized homeless people demanded that the sponsoring councilmember tell them where they could sleep if not in the parks of the city. The councilmember responded, "Why not City Hall?" About 100 homeless people—single men and women, families and the elderly—moved onto the City Hall lawn for 2½ months until the city agreed not to enforce the sleeping ban, essentially admitting that the legal control of public space rendered life impossible for homeless people (Howland 1994, 34–35).[15]

Yet despite, and quite likely because of, such protests, the legal exclusion of homeless people from public space (or at least the legal exclusion of behaviors that make it possible for homeless people to sur-vive) has increased in strength during the 1990s, creating and reinforc-ing what Mike Davis (1991) has called for Los Angeles "a logic like Hell's." This hellish logic is, of course, a response to another quite hell-ish one: the logic of a globalized economy that is successful to the de-gree that people buy into the ideology that makes their places to be little more than mere factors of production, factors played off other factors in pursuit of a continual spatial fix (Harvey 1982) to ever progressive cri-ses of accumulation. It is a response, then, that seeks to reregulate the spaces of the city so as to eliminate people quite literally made redun-dant by the very capital the cities now so desperately seek to attract.

It might seem absurd to argue that the proliferation of anti-homeless legislation is part of continual experimentation in devising a new "mode of regulation" for the realities of post-fordist accumulation (cf. Lipietz 1986). After all, the disorder of urban streets seems to bespeak precisely the inability to regulate the contemporary political economy. But, as Lipietz (1986, 19) argues, a "regime of accumulation" materializes in "the form of norms, habits, laws, regulating networks, and so on that ensure the unity of the process, i.e. the appropriate consistency of indi-vidual behaviors with the schema of reproduction"; and, as Harvey (1989, 122) further comments, such talk of regulation "focuses our attention on the complex interrelations, habits, political practices, and cultural forms that allow a highly dynamic, and consequently unstable, capitalist system to acquire a semblance of order to function coherently at least for a certain period of time." Hence, cities are grappling with two—perhaps contradictory—processes. On the one hand, they must seek to attract capital seemingly unfettered by the sorts of locational determinants important during the era when fordism was in development. That is, they must make themselves attractive to capital—large and small—that has the luxury of choosing one location among the many proffered. On the other hand, the cities (together with other scales of the state) must create a set of "norms, habits, laws, regulating networks" that legitimizes the new rules of capital accumulation, rules in which not only is location up for grabs but also companies seek returns of greater relative surplus value by laying off tens of thousands of workers in a single shot, outsourcing much labor or resorting to temporary employment agencies—that is, in which the creation of a reserve army of labor is seen as a positive good.

These two processes—making the city attractive to capital and encouraging the formation of a reserve army labor—are in many (but not all) ways contradictory, and they are continually negotiated within the urban landscape itself. Within capitalist systems, the built environment acts as a sink for investments at times of overaccumulation in the "primary" circuit of capital, the productive system (Harvey 1982, Ch. 8). This statement, however, should not be read to imply either that the landscapes thus produced are somehow "useless" to capital or that local elites, growth coalitions, or a more nebulous "local culture" has no direct influence on the form and location of such investment (see D. Wilson 1991). Rather, investment in the built environment is cyclical, occuring within an already developed built environment. "At any one moment the built environment appears a palimpsest of landscapes fashioned according to the dictates of different modes of production at different stages of their historical development" (Harvey 1982, 233). The key point, however, is that under capitalism this built environment must "assume a commodity form" (Harvey 1982, 233). That is, while the use values incorporated in any landscape may (for different parts of the population) remain quite important, the determining factor of a landscape's usefulness is its exchange value. Buildings, blocks, neighborhoods and districts can all be subject, as market conditions change, as capital continues its search for a "spatial fix," as other areas become more attractive for development, to rapid devaluation. Quoting Marx, Harvey (1982, 237) argues that "[c]apital in general is 'indifferent to every specific form of use value' and seeks to 'adopt or shed any of them as equivalent incarnations.' " People feel this in their bones; they understand the incredibly unstable, tenuous nature of investment fixed in immovable buildings, roads, parks, stores, and factories. If, therefore, the built environment appears as "the domination of past 'dead' labour (embodied capital) over living labour in the work process" (Harvey 1982, 237), then the goal of those whose investments are securely tied to the dead is to assure that the landscape always remains a living memory, a memory that still living capital finds attractive and worth keeping alive itself. Investments—dead labor—must therefore be protected at all costs.[16] If a built environment possesses use value to homeless people (for sleeping, for bathing, for panhandling) but that use value threatens what exchange value may still exist, or may be created, then these use values must be shed. The goal for cities in the 1990s has been to experiment with new modes of regulation over the bodies and actions of the homeless in the rather desperate hope that this will maintain or enhance the exchangeability of the urban landscape in the global economy of largely equivalent places. The annihilation of space by law, therefore, is actually an attempt to prevent those very spaces from being "creatively destroyed" by the continual and ever revolutionary circuits of capital.[17]

Hence, what cities are attempting is not a tried-and-true set of regulatory practices but rather a set of experiments designed to negotiate the insecure spaces of accumulation and legitimation at the dawn of the 21st century. The goal is to create, through a series of laws and ideological constructions (concerning, for example, who the homeless "really" are), a legitimate stay against the insecurity of flexible capital accumulation. That is, through these laws and other means, cities seek to use a seemingly stable, ordered urban landscape as a positive inducement to continued investment and to maintain the viability of current investment in core areas (by showing merchants, for example, that they are doing something to keep shoppers coming downtown). In this sense, anti-homeless legislation is reactionary in the most basic sense. As a reaction to the changed conditions of capital accumulation, conditions themselves that actively (if not exclusively) produce homelessness (see Marcuse 1988), such legislation seeks to bolster the built environment against the ever possible specter of decline and obsolescence. It actually does not matter that much if this is how capital "really" works; it is enough that those in positions of power and influence believe this is how capital works.[18] As Seattle City Attorney Mark Sidran told the city council, the purpose of stringent controls on the behavior of homeless people is "to preserve the economic viability of Seattle's commercial districts" (*Seattle Times* 1993b); or, as he wrote more colorfully in an op-ed piece, "we Seattleites have this anxiety, this nagging suspicion that despite the mountains and the Sound and the smugness about all our advantages, maybe, just maybe we are pretty much like those other big American cities, 'back East' as we used to say when I was a kid and before California joined the list of 'formerly great places to live' " (Sidran 1993). The purpose, then, is certainly not to gain hold of the conditions that produce so much anxiety. Regulation is designed not to regulate the economy but to regulate its victims.

Regulation is thus always ideological—a means of *displacing* scrutiny and blame. Indeed, regulating the poor (Pivin and Cloward 1992) has long been a primary ideological function of the state at both local and national scales. Such regulation is necessary, as Piven and Cloward (1992) show, because it is the means by which wages and other "drains" on capital accumulation may be minimized; it is how the state seeks to safeguard accumulation—and to maintain its own legitimacy by dividing factions of the exploited classes from one another. That we are in the midst of an ugly class war, centered on the "structural adjustment" of the welfare state and the criminalization of poverty, is certainly no news. But, beginning at least with the recession of the early 1980s, what does seem novel is the ferocity with which this goal is pursued: the rapid rise of the "revanchism" that Smith (N. Smith 1996, 1998) so compellingly details. Such revanchism as regards the homeless has worked in two steps. First there has been a reinvestment in a language of deviance and individual disorder at the expense of structural explanations for (and solutions to) the problem of homelessness. This accomplished, the second step has been to find the means to regulate—through law—this deviance and disorder, completing the turn away from any sense that homelessness might have extraindividual causes. The history of this shift in thinking about homelessness is worth briefly reviewing.

During the relatively stable long-term boom from the end of World War II until the early 1970s, homelessness in American cities was scripted quite clearly by discourses centered on deviance, disaffiliation, and alcoholism. The stereotypical homeless person was a single white male skid row bum subsisting on mission charity and fortified wine.[19] Considered misfits, wasted humans incapable because of their personal problems of realizing any part in the affluence the postwar period guaranteed to all those who wanted it, they were perhaps to be pitied, certainly to be shooed away from downtown, and carefully confined within traditional skidrows or other districts that had served the casual labor markets of the first half of the century.

The explosion of homelessness, and especially the "discovery" that women, children, and whole families were part of the homeless population in the 1970s and 1980s, brought with it the beginning of a change in discourses on homelessness. While the language of disaffiliation and deviance retained a certain prominence, homeless advocates worked hard to emphasize the structural determinants of homelessness (economic decline; the dismantling of the welfare state, of which deinstitutionalization can be seen as a part;[20] gentrification and redevelopment in areas susceptible to it; etc.).[21] This change in the tenor of the debate, however, was quickly met with a reassertion of claims about homelessness as an individual problem, claims that explicitly sought to turn debate away from economic causes. Perhaps the clearest sustained example of this reassertion of personal disorder as the primary cause of homelessness is Baum and Burnes's (1993) *A Nation in Denial*, which argues that not until we admit that the problem of homelessness is located within addicted and mentally ill individuals can we understand that structural explanations have done more harm than good. Greeted by a great sigh of relief by much of the media (cf. Raspberry 1992; Hamill 1993; Leo 1993), Baum and Burnes's argument can be seen as a primary plank in legitimizing the recrminalization of homeless people's behaviors (even if that was not the intent of the authors). Hamill (1993), for example, uses *A Nation in Denial* as a springboard for advocating "quarantining" homeless people on closed military bases.

Should anti-homeless legislation succeed, Hamill's "solution" will be redundant. The proliferation of anti-homeless legislation clearly indicates that the battle has largely been won by those who seek to repersonalize homelessness. Such legislation is possible only in the absence of an understanding that homelessness has extrapersonal structural determinants. Or, more accurately, troublesome homelessness is seen to reside in those who refuse the numerous social services proffered to them to help them negotiate the conditions that make them homeless. Whether homelessness is structurally produced or not, this logic goes, people *remain* homeless by choice.

So, for example, in an article praising San Francisco's Matrix program (a set of initiatives designed to enforce "public order" and force homeless people into the tattered social services system), MacDonald (1995, 80) wrote that the "city's efforts to place people in shelter have proved disappointing." Over 2 months, Matrix enforcement teams[22] tried to distribute 3,820 vouchers for a night's stay in a church shelter for men. But "less than half the vouchers were taken, and only 678 actually used." MacDonald (1995, 80) found even more alarming the fact that of 3,000 general assistance (GA) recipients who claimed homelessness in San Francisco during the mid-1990s (and who received $345 a month), only 700 took advantage of a voluntary program "whereby GA recipients can turn over their checks to a non-profit housing advocacy group which arranges for a discounted room in a clean and city inspected single room occupancy hotel." In exchange, the homeless person is given a $65 allowance for the month for all other expenses. When numerous contacts with homeless GA recipients failed to increase participation in this program, San Francisco voters made receiving GA contingent upon proof of housing. If a rent receipt could not be produced, a GA recipient would be offered shelter under the "volunteer" program. If the recipient refused, she or he would be stricken from the relief rolls. MacDonald's conclusion?

> In passing this measure, San Franciscans acknowledged that providing more housing and other services will be unavailing unless society no longer allows the utilization of those resources to be optional. Funding such services is, in any case, often irrelevant to achieving greater civility in the streets. Matrix has made an enormous difference in San Francisco, though it has placed few people in permanent housing. This suggests that merely enforcing long-standing norms of public conduct may have far more effect on reducing disorder than any number of social programs. (MacDonald 1995, 80)

Note the shift in logic here. No matter what the cause of homelessness, homeless people refuse to take advantage of all that society offers them. In that sense they *are* voluntarily homeless, and thus disciplining them is not only desirable but also necessary. Successfully reducing homelessness to a "lifestyle choice," MacDonald legitimizes all manner of punitive measures against those who "choose" it. "San Francisco is both a symbol of the past and the wave of the future. Pursuing freedom it got chaos. It is now re-discovering that liberty consists not in overturning social rules, but in mutual adherence to them" (MacDonald 1995, 80). As with all "little Arnolds," MacDonald fails to raise the question of who establishes these rules and who they serve (much less the question of how one is to live in San Francisco on $65 a month without panhandling); the implication that poor, homeless people have no *right* to the city could not be clearer. As Waldron (1991, 324) so clearly shows, "what we are dealing with here is not just 'the problem of homelessness,' but a million or more *persons* whose activity and dignity and freedom are at stake." But so too are we creating, through these laws and the discourses that surround them, a public sphere for all of us that is just as brutal as the economy that spawned the conditions in which homelessness developed.

CITIZENSHIP IN THE SPACES OF THE CITY: A BRUTAL PUBLIC SPHERE

Now one question we face as a society—a broad question of justice and social policy—is whether we are willing to tolerate an economic system in which large numbers of people are homeless. Since the answer is evidently, "Yes," the question that remains is whether we are willing to allow those who are in this predicament to act as free agents, looking after their own needs, in public places—the only space available to them. It is a deeply frightening fact about the modern United States that those who *have* homes and jobs are willing to answer "yes" to the first question and "no" to the second. (Waldron 1991, 304)

In the decade since Jeremy Waldron wrote these words, the crisis of homelessness in the United States has only deepened, and the vigor with which those of us with homes who answer "no" to Waldron's second question has only increased. But we often fail to realize the degree to which this "no"—and its codification in anti-homeless laws—is creating a truly *brutal* public sphere in which not only is it excusable to destroy the lives of homeless people but also there seems to be scant possibility for a political discourse concerning the nature and types of cities we want to build.[23] That is, anti-homeless laws reflect a changing conception of citizenship which, contrary to the hard-won inclusions in the public sphere that marked the civil rights, women's, and labor movements in past decades, now seeks to reestablish exclusionary citizenship as just and good.

Craig Calhoun (1992, 40) has argued that the most valuable aspect of Habermas's *The Structural Transformation of the Public Sphere* (1989) is that it shows "how a determinate set of sociohistorical conditions gave rise to ideals they could not fulfill" and how this space between ideal and reality might hopefully "provide motivation for the progressive transformation of these conditions." In later work, Habermas turned away from such an historically specific critique to focus on "universal characteristics of communication" (Calhoun 1992, 40). Others, however, have retained the ideal of a critical public sphere in which continual struggle seeks to force the material conditions of public life ever closer to the normative ideal of inclusiveness (as we saw in Chapter 1). Calhoun (1992, 37) suggests that social movements, not just dispassionate individuals, have been central in "reorienting the agenda of public discourse and bringing new issues to the fore" (see also Fraser 1990). As Calhoun (1992, 37) notes, the "routine rational–critical discourse of the public sphere cannot be about everything at once. Some structuring of attention, imposed by dominant ideology, hegemonic powers, or social movements, must always exist." Theories of the public sphere—and practices within it—therefore must always be linked to theories of public space (see Howell 1993). The regulation of public space necessarily regulates the nature of public debate: the sorts of actions that can be considered legitimate, the role of various groups as members of the legitimate public, and so forth. Regulating public space (and the people who live in it) "structures attention" toward some issues and away from others.

Similarly, the perhaps inchoate interventions into public debate made by homeless people through their mere presence in public forces attention on the private bodies and lives of homeless people themselves. This is the "crucial *where*" question to which Tim Cresswell (1996) has drawn our attention in his studies of social transgression. Cresswell argues that regulating people is often a project defined by the attempt to "purify" space, by the attempt to create for any space a set of determinant meanings as to what is proper and acceptable. Yet these proprietary rules are continually transgressed; and these transgressions are just as continually redressed through dominant discourse that seeks to reinforce the "network or web of meanings" of place such that the pure and proper is shored up against transgression. The object of such discourse,

Cresswell (1996, 59) writes, "is an alleged transgression, an activity that is deemed 'out of place' "—for example, just those sorts of "private" activities of the homeless in public space (see Staeheli 1996) that are now the subject of such intense legal regulation. By being out of place, homeless people threaten the "proper" meaning of place.

But there is more to it than that. By being out of place, by doing private things in public space, homeless people threaten not just the space itself but also the very ideals upon which we have constructed our rather fragile notions of legitimate citizenship. Homeless people scare us: they threaten the ideological construction that declares that publicity—and action in public space—must be voluntary (see Chapter 4). Efforts such as Heather MacDonald's (1995) to show the voluntary nature of homelessness are therefore crucial for another reason than that suggested above. Such efforts provide an ideological grounding for reasserting the privileges of citizenship, for assuring ourselves that our democracy still works, despite the unsettling shifting of scales associated with the annihilating economy. As homelessness grows concomitantly with the globalization of the economy (eroding boundaries, unsettling place, throwing into disarray settled notions about home, community, nation, and citizenship), homeless people marooned in public frighten us even more. Not there but for the grace of God, but rather there but for the grace of downsizing, outsourcing corporations, go I. So it becomes vital that we reorder our cities in such a way that homelessness is "neutralized" (Marcuse 1988) and the legitimacy of the state, and indeed our own sense of agency, is maintained. The rights of homeless people do not matter (when in competition with "our" rights to order, comfort, places for relaxation, recreation, and unfettered shopping) simply because we work hard to convince ourselves that homeless people are not really citizens in the sense of free agents with sovereignty over their own actions.[21] Anti-homeless legislation helps institutionalize this conviction by assuring that the homeless have no place in public to be sovereign.

Anti-homeless legislation, by seeking to annihilate the spaces in which homeless people must live—by seeking, that is, to so regulate the public space of the city that there is literally no room for homeless people—re-creates the public sphere as intentionally exclusive, as a sphere in which the legitimate public only includes those who (as Waldron would put it) have a place governed by private property rules to call their own. Landed property thus again becomes a prerequisite for legitimate citizenship. Denied sovereignty, homeless people are reduced to the status of children: "the homeless person is utterly and at all times at the mercy of others" (Waldron 1991, 229). Reasserting the child-like nature of some members of society so as to render them impotent is, of course, an old move long practiced against women, African Americans, Asians and some European immigrants, and unpropertied radical workers throughout the course of American history.

But such moves are not just damaging to their subjects. Rather, they directly affect the rest of us too. "[I]f we value autonomy," Waldron (1991, 320) argues,

> [w]e should regard the satisfaction of its preconditions as a matter of importance; otherwise, our values simply ring hollow so far as real people are concerned. . . . [T]hough we say there is nothing dignified about sleeping or urinating, there is certainly something inherently undignified by being prevented from doing so. Every torturer knows this: to break the human spirit, focus the mind of the victim through petty restrictions pitilessly imposed on the banal necessities of life. We should be ashamed that we have allowed our laws of public and private property to reduce a million or more of our citizens to something like this level of degradation.

We are re-creating a society—and public life—on the model of the torturer, swerving wildly between paternalistic interest in the lives of our subjects and their structured degradation. In essence, we are re-creating a public sphere that consists in unfreedom and torture. Or, as Mike Davis (1990, 234) puts it in a chillingly accurate metaphor: "The cold war on the streets of Downtown is ever escalating." To the degree we can convince ourselves that the homeless are the communists of our age, we are calling this public sphere right and just. And that has the effect of legitimizing not only our own restrictions on the autonomy of others but also the iniquitous political economy that creates the conditions within which we take such decisions.

LANDSCAPE OR PUBLIC SPACE?

> Building a city depends on how people combine the traditional economic factors of land, labor, and capital. But it also depends on how they manipulate symbolic languages of exclusion and entitlement. The look and feel of cities reflect decisions about what—and who—should be visible and what should not, [about] concepts of order and disorder, and [about] uses of aesthetic power. (Zukin 1995, 7)

The relationship between the annihilating economy and the annihilation of space by law made visible through anti-homeless legislation
is clearest in discourses on what could broadly be considered "aesthetics." When Senator Patty Murray (D-WA) was first elected to Congress
in 1992, she was shocked by what she saw on the streets of Washington,
DC: "I look around and see a city in shambles. . . . I see people in the
streets with cups next to me, and as I come to stop signs, begging for
money" (*Washington Post* 1993b). After quoting Senator Murray, the
Washington Post reporter continued: "The beggars, many but not all of
whom are homeless, are among those sights in the nation's capital that
tourists don't enjoy." And the executive vice president of the DC Convention and Visitors Association, Dan Mobley, added: "Panhandlers are
not a pretty picture." He also noted, however, that "we have never had
anyone say they won't come here because of the panhandlers" (*Washington Post*, 1993b). Even so, in numerous cities around the country,
concern with removing homeless people so as to restore the "pretty picture" remains a paramount obsession.[25]

The executive officer of Downtown Cincinnati (a business association), for example, has argued that "Panhandling today prevents many
visitors—from Cincinnati's suburbs and from out of town—from expe-
riencing and enjoying our beautiful downtown" (*Cincinnati Enquirer*
1995b). In Akron, a law criminalizing any (not just "aggressive") panhandling was supported by the mayor because "the city was trying to
clean up its downtown image with the opening of the new . . . Convention Center and the expected opening next year of Inventure Place, the
home of the Inventors Hall of Fame" (*Cleveland Plain Dealer* 1994).
And the interim president of the Atlanta Convention and Visitors Bu-
reau (ACVB) supported a comprehensive "crackdown on vagrants,
thugs and general trespassing": "I urge the ACVB and the community to
not help Atlanta" (*Atlanta Journal and Constitution* 1991). "We would
like as many tools as possible to keep the city clean," concurred San
Diego Police Captain George Saldamando (Rodgers 1992). Indeed, by
2000, the city of San Diego had turned the cleaning of streets—of both
litter and homeless people—over to a private program sponsored by the
Centre City Development Corporation and the downtown Property-
based Business Improvement District named "Clean and Safe" (Mitchell
and Staeheli 2002). As described by an official of the Downtown Partnership, which manages the program, "Clean and Safe" uses power-
hoses to wash the sidewalks of inanimate waste and "ambassadors" to

"get in the face" of homeless people and convince them to move out of the parks and off the sidewalks of downtown (quoted in Mitchell and Staeheli 2002). By doing so, the city, property owners, and merchants are convinced that tourists and middle-class suburbanites will find the downtown attractive and want to spend more time there.

In each of these instances the concern is with the appearance of the built environment of the city, with creating a landscape that does not "leav[e] a bad impression on visitors by feeding the impression that our downtown is unsafe" (*Cincinnati Enquirer* 1995a). The preferred method for doing this—the promulgation of anti-homeless laws (and in many instances turning their enforcement over to private security forces)—in essence seeks to re-create downtown streets *as a landscape*. The point I am making revolves around a particular definition of "landscape." As Denis Cosgrove (1984, 1985, 1993), Stephen Daniels (1993), and others (cf. Schein 1997) have shown so well, "landscape" implies a particular way of seeing the world, one in which order and control over surroundings takes precedence over the messy realities of everyday life. A landscape is a "scene" in which the propertied classes express "possession" of the land and their control over the social relations within it. A landscape in this sense is a place of comfort and relaxation, perhaps of leisurely consumption, unsullied by images of work, poverty, or social strife. Landscape, Cosgrove (1985, 49) shows, developed from and reinforces a "bourgeois rationalist conception of the world." More recently, Daniels and Cosgrove (1993; see also Cosgrove 1990, 1993) have explored the ways in which the landscapes operate not just as text, or as visual representation, but as the "theater" or stage upon which the "dramas" of life are enacted.[26] Yet, the sort of stage being constructed through the redevelopment of downtowns and their protection through anti-homeless laws is, like the festival marketplace or mega-mall that serve as its models, a theater for a "pacified public," as Crilley (1993) puts it (see Chapter 4), and as such it stages a spectacle in which the homeless have little or no part to play. Indeed, homeless people's constant intrusions onto the stages of the city seems to threaten the carefully constructed suspension of disbelief on the part of the "audience" that all theatrical performances demand, thereby seemingly turning that audience away and toward other entertainments: the suburban mall or the theme park (Sorkin 1992).

Anti-homeless laws are thus an intervention in urban aesthetics, in debates over the look and form of the city. "Aesthetic judgments,"

Harvey (1990, 429) has written, "have frequently entered in as powerful criteria of political and social action."[27] When these aesthetic judgments have the effect of valuing the spaces of the city as landscape rather than public space, they serve up a double "suspension of disbelief":

> The power of a landscape does not derive from the fact that it offers itself as spectacle, but rather from the fact that, as mirror and mirage, it presents any susceptible viewer with an image at once true and false of the creative capacity which the subject (or Ego) is able, during a moment of marvelous self-deception, to claim as his own. A landscape also has the seductive power of all *pictures*, and this is especially true of an urban landscape—Venice, for example—that can impose itself immediately as a *work*. Whence the archetypal touristic delusion of being a participant in such a work, and of understanding it completely, even though the tourist merely passes through a country or countryside and absorbs its image in a quite passive way. The work in its concrete reality, its products, and the productive activity involved are all thus obscured and indeed consigned to oblivion. (Lefebvre 1991, 189, emphasis in original)

Creating a city as a landscape therefore is important because it restores to the viewer (the tourist, the suburban visitor, or even the housed resident) an essential sense of control within a built environment, which is rather "controlled" through the creative, seemingly anarchic, destruction of an economy (operating at all scales) that can just as easily destroy the careers and lives of the viewer as it has already the people "downsized":

> Even in boom times, downtown Dallas was no field of dreams. In the early 1980s developers built it—stacking glass, steel and masonry ever skyward—but the people did not come. . . . Too soon, boom times departed as well. The corporate merger and acquisition phase that followed was marked by downsizing and consolidations that caused the office vacancies to skyrocket. Downtown Big D became the Big Empty. Decay followed. (*Houston Chronicle* 1995)

With the promulgation of anti-homeless laws, "as absolute political space extends its sway" in the name of safeguarding urban accumulation, "the impression of transparency" inherent in the landscape "becomes stronger and stronger" (Lefebvre 1991, 189).

If the illusion of control is one aspect of making over a city as landscape (through the "privatization" of public space that accompanies laws such as those directed at the homeless), then a second aspect is the reinforcing of an ideology of comfort, or what Sennett (1994, 18) has called the "freedom from resistance." To extend Sennett's argument, the urban landscape is increasingly designed not just to facilitate the movement of capital but also so that "citizens" can "move without obstruction, effort, or engagement" (Sennett 1994, 18).[28] "This desire to free the body from resistance," Sennett (1994, 18) argues, "is coupled with the fear of touching, a fear made evident in modern urban design." It is made even more evident in debates surrounding "aggressive panhandling" laws. The Washington, DC, begging ordinance passed in 1993 is typical. It prohibits "approaching, speaking to or following . . . in a way that would cause an ordinary person to fear bodily harm" (*Washington Post* 1993a; *Roll Call* 1993). Assault, of course, is already against the law, as is threatening harm. This law criminalizes not assault or threat-making but rather making someone feel uncomfortable. And panhandling, sleeping in public parks, or urinating in alleys makes us, myself included, necessarily uncomfortable. As it should. Discomfort, however, is a far cry from either "wrong" or "dangerous," even if we are frequently reluctant to make such distinctions. "He said, 'I want you to do me a favor.' I said, 'I don't have any money.' I figured that is what he wanted. It really scares me," an elderly woman in Memphis reported. "I don't have a gun, but this is one time [I wish I did]" (*The Commercial Appeal* (Memphis), 1994). The fear of bodily contact is often less palpable than that expressed by this woman,[29] but it shows up in concerns over our ability to move down a street or into a place of business without encountering a homeless person. "The city street gauntlet may include six panhandlers in one block. A few sit silently on a bench or crouch against buildings, thrusting plastic-foam cups at the strangers who rush past them. Most, either through their signs or pleas, make more direct requests for money" (*Washington Post* 1993b). Even the most passive of beggars are threatening: the street becomes a "gauntlet" and the silent continually "thrust" toward you.

Sennett argues that "the ability to move anywhere, to move without obstruction, to circulate freely, a freedom greatest in an empty volume" has come to be defined as freedom itself in "Western civilization."

> The mechanics of movement has invaded a wide swath of modern experience—experience which treats social, environmental, or personal resistance, with its concomitant frustrations, as somehow unfair and unjust.

Ease, comfort, "user-friendliness" in human relations come to appear as guarantees of individual freedom of action. (Sennett 1994, 310)[30]

There are two important points here. First, such freedom of movement is only possible by denying others the same right (cf. Blomley 1994a, 1994b). Anti-homeless laws have been challenged on the grounds that, by effectively banning some people from public spaces, they are in violation of homeless peoples' constitutional right to travel (Ades 1989; Simon 1995; Mitchell 1998b). Hence, *our* mobility is predicated on the immobility of the homeless. The homeless provide "resistance" to our unfettered movement, cause discomfort as we try to navigate the city. And those homeless people who persist in challenging our right to walk by without helping them to survive are anything but "user-friendly."

The second point is that this ideology of comfort and individual movement as freedom reinforces the "impression of transparency" that works to make the urban landscape knowable by erasing its "products and productive activity." "[R]esistance is a fundamental necessity of the human body," Sennett (1994, 310) concludes: "Through feeling resistance, the body is roused to take note of the world in which it lives. This is the secular version of the lesson of exile from the Garden. The body comes to life when coping with difficulty." Reflecting on the construction of a city built on the ideal of the *flâneur* in the 19th century, Sennett (1994, 347) further argues that "a public realm filled with moving and spectating individuals . . . no longer represented a political domain." And in places like contemporary Greenwich Village (where Sennett lives) or other urban neighborhoods, "ours is a purely visible agora" where "political occasions do not translate into everyday practice on the streets; they do little, moreover, to compound the multiple cultures of the city into common purposes" (Sennett 1994, 358). This, of course, is ever more the case as city government after city government seeks to enhance city images by engaging in "quality of life" campaigns.

In short, "quality of life" initiatives such as anti-homeless laws raise a politics of aesthetics above the politics of survival. They substitute an image of the urban landscape for a grounded politics of place designed to improve the lives of all the people of the city. They reduce the "right to the city" for all to a "right" for some to be free from the annoying "resistance" of those thrown into the streets they want to walk on. Crilley (1993, 157) sets "megastructures" like Canary Wharf in London or the World Financial Center in New York—structures fully controlled such that they reproduce the life of the city as nostalgia—against the "traditional city." "Traditional cities," he writes, "with their connotations of vitality, social interaction and heterogeneity, cannot be 'programmed' or 'animated'; history and memory in the city do not have 'essences' reducible to visual images; and a genuine public presence cannot be engineered through the application of correct forms, dazzling spectacle, or the lure of free bread and circuses." Yet, this is precisely what cities are attempting with the crackdown on homeless people. They seek to replace the public spaces of the city with landscape, to substitute the visual for the (often uncomfortable and troublesome) heterogeneous interactions of urban life.

If malls and festival marketplaces represent one pole of what Michael Walzer (1986) has called "closed-minded" public spaces (those spaces designed for a single function, spending at the expense of hanging out, for example, or, better yet, hanging out as a means to induce spending), then anti-homeless laws represent the other pole. "In 1994, the message in many U.S. cities to people on the street," noted columnist Colman McCarthy (1994), "was either get lost or get arrested." Things have only deteriorated since then (Foscarinis, Cunningham-Bowers, and Brown 1999).

CONCLUSION

Public space—like the right to the city—is always a negotiation (see Goheen 1993). The proliferation of anti-homeless laws ups the ante in these negotiations by seeking explicity—and within the realm of law—to remove some people from the negotiators' table through the simple expedient of turning them into criminals. These laws have as a goal—perhaps not explicit, but clear nonetheless—the redefinition of public rights so that only the housed may have access to them. They further the goal of redefining the public space of the city as a *landscape*, as a privatized view suitable only for the passive gaze of the privileged as they go about the work of convincing themselves that what they see is simply natural.

The genealogy of these laws in the insecurity the contemporary bourgeoisie feels within the "globalizing" economy seems clear enough. In an era in which the "symbolic economy" has risen to replace a seemingly more stable industrial-based economy, the "culture of cities" is everything (Zukin 1995). This "culture"—this landscape—is itself a tenuous thing, not at all a sure or permanent attraction to footloose capital. The rise, then, of what Zukin calls the "aestheticization of fear" seems a quite understandable, if still appalling, thing. By creating superficially pleasing landscapes we hope to stave off the inevitable, to steal from history a few more months or years of prosperity. If this genealogy is clear, however, so too are the costs. Anti-homeless laws are perhaps the clearest indication of the Faustian bargain we are daily making to protect our own relative affluence. The cost to homeless people we so willingly sacrifice is of course the greatest cost. But so too is the rather unthinking construction of a brutal public sphere a high price to pay for an attractive downtown. "Fear proves itself," Mike Davis (1990, 224) quotes William Whyte as saying, while adding himself that the "social perception of threat becomes a function of the security mobilization itself, not crime rates."[31]

Indeed, anti-homeless laws indicate the degree to which the public sphere, modeled as it is on the palpable fear of the bourgeoisie, has become less a place of critique, debate, and struggle—a place where the cry and demand for the right to the city is heard over and over again—and more an arena for the attempted legitimation of a brutal political economy and landscape as a just political economy and landscape.

NOTES

1. As we will see, the use of quality-of-life laws for such purposes is quite consistent with their overall purpose: the elimination of public space.
2. Cincinnati's ordinances, passed in 1995, were struck down in court. New ordinances, more narrowly drawn, but achieving the same thing were passed in 2002. See *Cincinnati Enquirer* (2002).
3. The original revanchism was a movement of reaction against both the royalty and the working class in late-19th-century France. Deeply nationalist, it mobilized around "traditional values" (Smith 1996, 45).
4. For an excellent analysis of this dynamic, see Henderson (1999, Ch. 2).
5. It is remarkable how often, now, investment in museums or concert halls (and even more so stadiums) is sold to the public not because it might make the city a better place to live but because it will make it "competitive"—a remarkably anemic reason for public investment.
6. In July 2001, the Los Angeles City Council voted to approve the installation of pay toilets around that city's skid row, so perhaps defecating in alleys will no longer be as great a need there. The provision of toilets was approved after a 20-year debate. Really (*New York Times* 2001).
7. Designed to control homeless people, these laws were found to be of great value during the 1999 protests at the WTO meeting in Seattle, when hun-

dreds of activists were arrested, not for rioting, but for such quality-of-life offenses as sleeping in a public place (Falit-Baiamonte 2000). The mutability of law—its ability to be transferred from one realm of control to another—is something we ignore to our great peril.

8. The reference is to the 1993 Meg Ryan and Tom Hanks movie *Sleepless in Seattle*, about a relationship formed in a cyberspace of sorts: a late-night coast-to-coast chat radio show.

9. This despite a protestation from the Assistant City Attorney that the law *did not* target panhandlers. The original *Seattle Times* article quoted above included the phrase "ordinances that would ban panhandlers from sitting on sidewalks." Later editions carried the following: "Correction: The City's sidewalk ordinance prohibits sitting or lying on sidewalks in business areas and does not target panhandlers according to Assistant City Attorney Laurie Mayfield. This article indicated otherwise." The law certainly did not target espresso sippers.

10. Shelters are no exception. Homeless people are required to behave according to the rules established by their operators, and their ability to remain in the shelter is at the sufferance of the management.

11. It is also why advocates for the homeless have sought to contest anti-homeless laws—so far without much success—on both "right to travel" (and stay put) and cruel and unusual punishment principles (Simon 1992; Mitchell 1998b).

12. As Robert Ellickson (1996) details, this is the direction a number of cities are being forced to move in by courts as they respond to the sorts of arguments that Waldron lays out.

13. As we will see in the next chapter, the urban right takes a very different view of freedom. According to legal scholars such as Ellickson (1996), it seems to consist only in not being jailed.

14. Of course, Engels made much the same point a century and a half ago in his examination of the condition of the working class in Manchester. He argued there that the bourgeoisie had no real solution to the housing problem except to move the poor about, shifting the crisis from one district to another. Too little has changed in the ensuing years.

15. Similar actions have been repeated elsewhere in California: see Johnson and Moroo (1996).

16. I develop a theory of landscape based on "dead labor" more fully in Mitchell (2001a) and (2003).

17. My point is not at all that the globalization of capital is some sort of *deus ex machina* over which we have no control. Rather, it is that a contradiction exists between the need for ever faster turnover times by capital in general and the need to fix some capital in particular places. Capital needs places. But the question is always one of *which* places, endowed with what sorts of attributes, and this is a question that is only answered in practice. That being so, people with investments rooted in particular places find their investments to be quite insecure. Property, a necessary condition of capital accumulation, can also be rapidly devalued, in essence mortgaging the suc-

cess of some kinds of investment against the loss of other kinds. Capital is not united, and its complex divisions and contradictions are precisely what lead to the overweaning sense of insecurity that governs most American cities.

18. As the Berkeley merchant's self-fulfilling argument quoted in the preceding chapter indicates.

19. For reviews and examples of discourses on homelessness (in chronological order), see Sollenberger (1911); Anderson (1923); Dees (1948); Bahr (1970, 1973); Spradley (1970); Blumberg, Shipley, and Barsky (1978); Hopper and Hamberg (1984); Schneider (1986); Hoch and Slayton (1989); Rossi (1989); Baum and Burnes (1993); Wolch and Dear (1993); Takahashi (1998).

20. This was not the only reason for deinstitutionalization, of course. Its history is much more complex than that and incorporates much that is good, such as the desire to dismantle "total institutions" for the physically and mentally ill.

21. Some examples include Hombs and Snyder (1982); Hopper and Hamberg (1984); Kasinitz (1986); Mair (1986); Dear and Wolch (1987); Hartman (1987); Marcuse (1988); Deutsche (1990); Blau (1992); Veness (1993); Wolch and Dear (1993). Good recent reviews are Takahashi (1996; 1998, 4–13).

22. The program has two components. First, police engage in something like "a military campaign. Retaking the city block by block. Every ten days or so, the Matrix teams would announce a sweep of an additional area chosen on the basis of citizen complaints." Second, "Matrix also included social service outreach. A team of two social workers, two mental health workers, a substance abuse specialist, and two police officers roams the city trying to coax the homeless into shelters, housing programs, or treatment for addiction and mental illness" (MacDonald 1995, 79).

23. This issue has certainly been raised in the wake of the September 11, 2001, terrorist attacks and is now subject to at least some debate. Jeffrey Rosen's (2001) article in the New York Times Magazine on closed-circuit television in Britain and its panoptic features is an important prominent intervention into the rising tide of discourse that takes "security" to be the primary issue at stake in public space. The problem with much of the discussion in the month immediately following the attack was that it, perhaps understandably, occurred in a vacuum. Rosen's article was one of the few that I have seen that drew, if only partially, on the years of research and debate about surveillance in public space and its relationship to freedom, politics, and the rights of the most vulnerable.

24. Let me be clear: the various ideologies through which we understand the homeless and homelessness are indeed contradictory. On the one hand, we need to show that homelessness is a voluntary rather than a structural condition. On the other hand, we also need to show that homeless people are not citizenly "free agents," a position that seemingly undermines the ideology of volunteerism. Yet, the contradiction is resolved quite simply: since

homeless people have chosen to be (or remain) homeless, they are therefore ineligible for legitimacy.
25. The paradigmatic accounts are Davis (1990) and Sorkin (1992).
26. For a fuller examination of the stage metaphor in landscape geography, see Mitchell (2000, Ch. 5).
27. This point is driven home with force for the case of upper-class suburbia outside New York by Duncan and Duncan (2001).
28. In this regard, those aspects of "bubble laws" (see Chapter 2) that establish a safe zone around individuals as they move through cities become doubly interesting: could it be that we are beginning to see the development of a legal regime that takes personal sovereignty as a state of legal isolation from all that one does not wish to encounter? It is not hard to imagine a world in which individuals are legally granted a "sovereign space" that moves with them through the city, keeping beggars, leafleters, and strangers at bay.
29. I am certainly not implying that the fear felt by this woman was not real. Rather, the question is whether our fear or discomfort should be allowed to dictate the destruction of the means of survival for other people. Is the drawing of a gun really an appropriate response to being panhandled?
30. The degree to which Sennett is describing a largely white, male, and bourgeois ideology should be obvious. Clearly the dream of a resistance-free public sphere for some—that is, a fully deracinated "right to the city" for the pampered classes—has been historically predicated on a dystopian nightmare for most.
31. Again, see the recent New York Times Magazine article by Rosen (2001) on this point.

6

No Right to the City

*Anti-Homeless Campaigns, Public Space
Zoning, and the Problem of Necessity*

To what degree are the political economy, landscape, and public sphere brutal? In Colorado, more than one homeless person dies—from exposure, assault and murder, lack of medical attention, being hit by a car, and so on—every week (*Denver Post* 2000b); one a week also dies in California's Santa Clara County (*San Francisco Chronicle* 2000). Rates of death vary considerably by city. Boston averaged about one death every 2 weeks during the 1980s and 1990s, but that dropped to only four deaths for all of 1999. City officials declared that a new concerted effort to reduce street deaths was responsible for the decline (*Boston Globe* 2000). Such efforts are immensely important but rare. In "liberal," warmer San Francisco, some 157 people died on the streets in 1998. In 1999, that number increased to 169, continuing an almost constant rise in homeless deaths throughout the 1990s—that is, throughout the period of San Francisco's much lauded Matrix program (*San Francisco Chronicle* 1998, 1999).[1] Such statistics are hard to come by. They are relatively abundant for San Francisco only because a small, social-action-oriented newspaper, the *Tenderloin Times*, began compiling police reports and news records. In other cities, such as Denver and Boston, homeless advocates try to keep track of deaths and every year sponsor a "homeless memorial day" to remember those who died on the streets, but in most American cities the number of homeless deaths is simply not tracked. There is no way of knowing just how many homeless people die on the streets each year, but if the numbers from Boston and San Francisco represent the range for big cities, then it must be several thousand.

But these deaths seem almost "accidental." Even the rash of murders of homeless people in Denver in 1999 (seven homeless people were killed, mostly beaten to death, in 3 months) seems simply to be a tragedy—one to be condemned but not one for which the state is culpable. Yet, state policy *is* crucial, for it is precisely America's housing and homeless policies—together with the regime of private property that such policy supports—that put homeless people at risk of murder, death by exposure, and so forth. As Laura Weir, policy director for the National Law Center on Homelessness and Poverty, correctly notes: "Living in public places makes them easy targets. Homeless are at increased risk for violence being committed against them" (*Denver Post* 1999b). As the number of homeless people around the United States continued to climb in the late 1990s, the average wait to be placed in a public housing unit increased from 10 to 11 months between 1996 and 1998, and the wait to receive housing vouchers grew from 26 to 28 months, according to Department of Housing and Urban Development numbers (*Denver Post* 1999b). Bed space in shelters in Denver, as elsewhere around the country, is woefully inadequate. In the mid 1990s, according to the Department of Housing and Urban Development, and by the *most conservative* estimates, there was a nationwide shortage of *at least* 425,000 beds each night in the shelter system (Foscarinis 1996). Yet, a revivified housing program (much less a more adequate shelter system) is not even close to being on the policy or political agenda. "Living in a public place" is simply the only option that thousands of people have. As Foscarinis (1996, 14) puts it, "The discrepancy between need and emergency aid means that each night at least 425,000 people have nowhere to sleep except in public places, and that each day at least 700,000 people . . . have nowhere to be save public spaces. At the very minimum this means that they must perform essential bodily functions—such as sleeping, eating, bathing, urinating and defecating—in public."

Instead of working toward a more just housing and shelter system in the United States, the official line is more geared toward demonizing homeless people—making homeless people seem somehow less than human, endowed with fewer rights than those of us who live in houses. If there has been an overriding discourse about homeless people over the past decade, it has been that they are nuisances (or worse) to be rid of—pests and vermin who sap the economic and social vitality of the cities and the nation. Consider this statement by New York Mayor Rudy Giuliani, made in a report announcing a new policing strategy for New York's public spaces aimed at ridding them of homeless people: "Disorder in the public space of the cities" presents "visible signs of a city out of control, a city that cannot protect its space or its children" (Giuliani and Bratton 1994, 5, as quoted in N. Smith 1998, 3). This statement suggests that our children, our very patrimony, are at risk, and the spaces of the city cannot be readily defended from the internal rot that is homelessness, a rot that must be eradicated. There is something officially organized, something deeply rooted in American urban and national policy, about the demonization of homelessness that makes their ongoing murder, death by exposure, or lack of medical care appear to be the result of their homelessness rather than the result of inadequate or faulty housing, mental health, drug, and employment policies.

In June 1988 the Santa Ana, California, parks director wrote in a memo to his staff that the "City Council has developed a policy that vagrants are no longer welcome in the city of Santa Ana. . . . In essence the mission of this program is to move all vagrants and their paraphernalia out . . . by continually removing them from the places that they are frequenting in the City."[2] That year, city police began a series of indiscriminate sweeps of the city's sidewalks, parks, and other public places, sweeps that included the wholesale confiscation of personal belongings. The Legal Aid Society and the American Civil Liberties Union began preparing a lawsuit against the city, but this did little to halt the street sweeps. Two years later, on August 15, 1990, city police decided to organize a "deportation" of homeless people from Santa Ana (the label "deportation" is the police chief's). Police arrested some 64 homeless people for various minor crimes such as jaywalking, public drunkenness, urinating in public, littering, and "picking leaves from a tree" (Eng 1991) and drove them to the unused Santa Ana municipal stadium. There they chained the arrestees to benches, some for as long as 6 hours, and wrote identification numbers on their bodies in indelible ink. Most of the detainees were cited for littering and eventually released; four were arrested on outstanding warrants; 19 Hispanic men, including at least one legal permanent resident, were handed over to the Naturalization and Immigration Service and deported to Tijuana (Gomez 1990; Simon 1995; Takahashi 1998). Despite objections from

civil rights, immigration, and legal aid attorneys that the Santa Ana police were engaging in "Nazi practices" (Gomez 1990), the chief of police, Paul Waters, promised to continue sweeping the homeless out of
the Santa Ana Civic Center and surrounding areas (Eng 1990a). He had
considerable support from the city council. As one member noted,
though he was upset that the police had not given him advance warning
of the sweep, he was still fully supportive: "My constituents would just
as soon wipe the slate clean of the homeless problem," he told the *Los
Angles Times* (Gomez 1990). "I know situations where there are truly
homeless people, but these are vagrants, bums and panhandlers. . . .
They don't truly want to help themselves. They absolutely don't want to
stop begging, stealing, and bumming around." And as a spokeswoman
for the Santa Ana Police Department noted, the sweep was necessary because of "a general rise in crime." She told the *Times* that "there have
been 86 thefts from cars and 22 stolen cars reported this year" (Gomez
1990). Though it was only homeless users of the Civic Center who were
rounded up (many of whom were Hispanic), the spokeswoman was
adamant that the sweep was "not directed against the homeless or the
Hispanic community."

Even so, when the sweeps were resumed (as promised) on August
21, once again only presumptively homeless people were rounded up
(Eng and Drummond 1990). Indeed, as they later admitted in a trial, in
the original sweep the police had released two of the detainees because
they could prove a fixed address (Eng 1991). On the second sweep, 26
people were arrested on misdemeanor charges. Some 18 officers, including several stationed on rooftops with high-powered binoculars,
took part in the operation, finding homeless people engaged in such actions as standing behind a tree with "a napkin or something in his
hands" (Eng and Drummond 1990). Together, the two sweeps detained
90 homeless people. The theory behind the detentions was that, while
not all "of the increase in crime is attributed to the homeless," as a police lieutenant said at a briefing, "we know that some of the homeless
are . . . committing some of the crimes." That is to say, homeless people
as a class were being rounded up because *some* homeless people may
have committed crimes. The appalling implications—in terms of basic
human rights, let alone the right to the city—are clear enough: whole
classes of people are being made suspect and their elimination is regarded as not only desirable but also socially necessary.[3]

The necessity of sweeping homeless people from public space was baldly stated by Police Chief Walters: while it was "unfortunate that a segment of our society has been driven to seek shelter" in the Civic Center, if they are allowed to remain, their presence "will not only lead to more serious crime but also certainly contribute to the belief that democratic government has become totally futile" because some members of the public are allowed to engage in activities that others—those who follow community "standards of behavior"—do not (Walters 1990, B9). Walters based his argument on the well-known "broken windows" thesis of criminologists James Q. Wilson and George F. Kelling (1982). For Wilson and Kelling, homeless people are little more than "broken windows" that signal the deterioration of community and the ready availability of a neighborhood for crime. Broken windows must be fixed if flourishing neighborhoods are to be maintained—or so goes the theory.

The "broken windows" theory is a particularly compelling and, at the same time, a particularly pernicious theory of public space. A major weapon in the ongoing war against homeless people, it has come to be taken, in many policy circles, as simple common sense (despite the fact that it probably does not work even on its own terms: see Harcourt 2001a), and it has served to license some quite remarkable experiments in the depletion of homeless people's rights, especially their right to some space in the city. In this chapter we will continue our examination of the ongoing war against homeless people in American cities by exploring a set of variations on public space zoning that the broken windows thesis and policing actions like those in Santa Ana have led to. The aftereffects of the Santa Ana roundups of the summer of 1990 are, in fact, quite complex, both legally and socially, and we will use the case of Santa Ana as a continual touchstone in this examination. What will become clear is that both in law and in practice American policy makers are continually seeking out new ways to make sure homeless people have no right to the city—even when the laws they construct turn out to be as constitutionally invalid as they are morally repugnant.

"BROKEN WINDOWS"

As indicated in Chapter 1, the problem of public space is often posited as a problem of order. Proponents of quality-of-life ordinances and policing argue that "disorder" is the primary threat facing urban neighborhoods. George Kelling, one of the authors of the broken windows thesis, writing with the legal scholar Catherine Coles, defines disorder as "incivility, boorish and threatening behavior that disturbs life, especially urban life" (Kelling and Coles 1996, 14). "By disorder," they continue,

> we refer specifically to aggressive panhandling, street prostitution, drunkenness and public drinking, menacing behavior, harassment, obstruction of streets and public spaces, vandalism and graffiti, public urination and defecation, unlicensed vending and peddling, unsolicited window washing of cars ("squeegeeing"), and other such acts. (p. 15)

Each of these may already be outlawed misdemeanors or petty offenses, they note, but they are not, in their estimation, policed stringently enough, either by the police themselves or less formally by neighborhood groups. This is particularly problematic, they argue, because the failure to guard against such disorder is the equivalent of leaving a broken window—or a whole neighborhood of broken windows—unmended.

James Wilson and George Kelling (1982) argue in their original statement of the thesis that even a single broken window in an urban neighborhood indicates a lack of care about urban space that invites other, more serious, criminal behavior. A single broken window, they argue (31), indicates that a building and surrounding property will "become fair game for people out for fun and plunder." Such a broken window is "criminogenic," as Kelling and Coles (1996, 15) term it. But the key to this argument is that "broken windows" are only a metaphor—and not for urban disinvestment. Rather, they are a metaphor for "disorderly behavior." Make no mistake, proponents of the broken windows thesis are very clear about what—or rather *who*—needs to be policed and subject to strict control. "The citizen who fears the ill-smelling drunk, the rowdy teenager, or the importuning beggar," Wilson and Kelling (1982, 29) write,

> is not merely expressing his distaste for unseemly behavior; he is also giving voice to a bit of folk wisdom that happens to be a correct generalization—namely, that serious crime flourishes in areas where disorderly behavior goes unchecked. *The unchecked panhandler is, in effect, the first broken window.* . . . If the neighborhood cannot keep a bothersome panhandler from annoying the passers-by, the thief may reason, it is even less likely to call the police and identify a potential mugger or to interfere if a mugging takes place. (emphasis added)

Later (35) they argue that "arresting a single drunk or a single vagrant who has *harmed no identifiable person* seems unjust . . . [but] failing to do anything about a score of drunks or a hundred vagrants may destroy an entire community" (emphasis added). That is to say, the more people there are *who are harming no identifiable person* but merely engaging in what the authors declare to be "disorderly behavior," the more just is the engagement in an unjust act, for "disorderly behavior" in and of itself poses a "grave threat . . . to our society" (Kelling and Coles 1996, 7). Indeed, Wilson (1996, xiv) has even formulated a new Malthusian law to describe this threat: "As the number of unconventional individuals increases arithmetically, the number of worrisome behaviors increases geometrically." So now the threat is not even disorder, but only "unconventionality," and the "harm" (though unidentifiable) is creating "worry,"[4] and on this basis authorizing the police to push the homeless along, giving Business Improvement Districts the power to "sweep" the streets of the homeless, and writing laws that make it illegal for certain individuals to sit, rest, sleep, or even eat are all justified.[5] "Broken windows," in short, is a policy of "zero tolerance" for behaviors and actions deemed disorderly or "worrisome."

Even if read generously, the logic of the "broken windows" thesis is incredible. "Untended *disorderly behavior*," Kelling (1987, emphasis in original) has written, "communicate[s] that nobody cares (or that nobody can or will do anything about the disorder) and thus [may] lead to increasingly aggressive criminal and dangerous predatory behavior." That is to say, avowedly innocent people need to be targeted by the police, the law, and the "community" because of the potential in a particular place for *other people* to commit crimes.[6] At its boldest and baldest, this defense of punitive measures against the homeless asserts that the *aesthetics of place* outweigh other considerations, such as the needs of homeless people to sleep, to eat, or to *be* (Waldron 1991).[7] As law scholar Steven Taisner (1994, 1272) argues in the midst of an attempt to develop constitutionally valid means of ridding city streets of homeless people, "the *most serious* of the attendant problems of homelessness is its devastating effect on a city's image" (emphasis added), and not, presumably, such attendant effects as ill health, mental illness, malnutrition, or death from exposure. Robert Tier (1998) of the American Alliance for Rights and Responsibilities and the Center for Livable Cities draws heavily on Kelling and Coles (1996) to make the same point, arguing that the presence of homeless people on city streets or in parks "affect[s] the quality of urban life, the general feeling of comfort, aesthetics, security, and freedom people should have in their urban public spaces" (Tier 1998, 258).

Advocates of the "broken windows" thesis readily admit that there are constitutional problems with singling out classes of people and criminalizing them for their class status.[8] It is impermissible, for example, to outlaw panhandlers as a class, though some forms of *panhandling* as a behavior can be made illegal. Anti-camping ordinances have been defended on the grounds that they outlaw a specific action—sleeping in public—rather than a status (being without a home). Opponents of such laws have contested them by arguing that homeless people *have no choice* but to sleep in public (or to trespass, which is illegal), that sleeping is necessary to survival, and hence anti-camping ordinances *effectively* target a status (homelessness).[9] Proponents of the broken windows thesis and other forms of public order policing counter that the *behaviors* they seek to criminalize are, in fact, *voluntary.*

George Kelling and Catherine Coles (1996), in fact, agree that homelessness is a *status* or *condition*. They write: "The problem is not the *condition* of being homeless or poor; it is the *behavior* of many persons, some homeless and others not, who violate the laws of the city and state" (40, emphasis in original).[10] They go on to argue that "using 'homeless' as a euphemism for a panorama of antisocial and/or unlawful activities, and for 'the poor,' gives all who are poor a bad name, and ignores the reality that most poor are law abiding, embody a sense of decency and respect for others, and take responsibility for their own obligations" (65). To make their argument, Kelling and Coles quote a particularly odious passage from Baum and Burnes's[11] (1993) analysis of the issues at stake in a celebrated court case in San Francisco concerning the rights of the homeless (*Joyce v. San Francisco* 1994), a passage that in fact undermines their very argument:

> By perpetuating the myth that the homeless are merely poor people in need of housing, . . . advocates reinforce and promote the most pernicious stereotypes about poverty in America. The vast majority of poor people in America are not homeless. Poor people do not live on the streets, under bridges, or in parks; do not carry all of their belongings in shopping carts or plastic bags; do not wear layers of tattered clothing and pass out or sleep in doorways; do not urinate or defecate in public places; do not sleep in their cars or encampments; do not harass or intimidate others; do not

ask for money on the streets; do not physically attack city workers and residents and do not wander the streets shouting at visions and voices. . . . (as quoted in Kelling and Coles 1996, 65)

On the one hand, then, Kelling and Coles admit that homelessness is a "status" or "condition," but on the other hand they rely on a definition of homelessness (the implication that Baum and Burnes are in fact defining homelessness in this passage is obvious enough) that sees it, a priori, as criminal or antisocial. The homeless by this definition are not law-abiding and are worthy of very little empathy. Indeed, in this telling, homelessness—or housing—per se is not the issue. In Baum and Burnes's (1993, 2) words, "emerging research" has indicated that perhaps as much as "85 percent of all homeless adults suffer from chronic alcoholism, drug addictions, mental illness, or some combination of the three, often complicated by serious mental problems" (quoted in Kelling and Coles 1996, 66). What neither Baum and Burnes nor Kelling and Coles go on to do is make the obvious point: just as the *majority* of poor people are not homeless, neither are the *majority* of mentally ill, drug-addicted, alcoholic, or otherwise disabled people homeless. Making that point would entirely undermine the argument that both sets of scholars are seeking to make—namely, that the issue at stake on the city streets is not homelessness, but behavior. For, if that further obvious point were made, then we would have to admit that what is at stake is precisely *homeless*ness, even if this is indeed linked to numerous other social and personal problems.[12]

Nonetheless, assuming that behaviors associated with homelessness are voluntary pays dividends: it allows for such behavior to be, or to remain, criminalized. Making this assumption further allows Kelling and Coles (222), and by extension those jurisdictions that adopt "order-maintenance" policing, to claim that "although order maintenance activities will put police into contact with homeless and poor people, those who are emotionally disturbed, youths, and substance abusers, order-maintenance efforts are not intended to solve society's problems regarding these populations." There are two problems with this argument (besides its corollary demonization of all "youths"), and with broken windows policing targeted at street people more generally. First, the distinction between status and behavior is often a false one *by necessity*. The second problem is that broken windows policing has de facto become the *only* attempted solution to the problem of homelessness in the

United States. Returning to the streets and courthouses of Santa Ana will allow us to explore both these points in more detail.

SANTA ANA'S ANTI-CAMPING ORDINANCE AND THE PROBLEM OF NECESSITY

Santa Ana's Police Chief Paul Walters (1990) saw his roundup of homeless people as, in his words, a means of "fixing [the] public's broken windows," but others were deeply appalled by the action. Almost immediately, the Legal Aid Society and the American Civil Liberties Union sued to halt the police department from indiscriminately detaining and "deporting" homeless people from the Civic Center. It was not their first suit against the city. In 1988 there were probably between 5,000 and 6,000 homeless people in Orange County, of whom perhaps half lived in the downtown Santa Ana area (Schwartz and Kurtzman 1988, 13, cited in Takahashi 1998, 156).[13] Santa Ana is the county seat, and thus numerous social service agencies are located there. Throughout the 1980s Santa Ana was engaged in redeveloping its Civic Center area. This, coupled with the sense that the municipality of Santa Ana was shouldering a disproportionate share of the burden for Orange County's homeless, led the city government to institute a series of police sweeps in which homeless people were either arrested or "moved along" and their belongings were confiscated and destroyed (Takahashi 1998, 156–157). The ACLU and the Legal Aid Society, representing affected individuals, sued, and in early 1990 won a settlement from the city awarding 17 of those affected by the sweeps $50,000 each. Homeless people, according to the settlement, would still be subject to sweeps, but now their belongings would be stored as lost-and-found items rather than destroyed (Eng 1990a).

Six months later the city of Santa Ana made good on its promise to continue making sweeps by embarking on its infamous deportation. In addition to filing a new suit, the Legal Aid Society and the ACLU promised to seek a jury trial for each of the individuals arrested. They received support in this action from the bar association (Eng 1990b). By October 1990, attorneys from the Legal Aid Society of Orange County and the Orange County Bar Association had secured another agreement from the city. Among its 22 points, the agreement "specifically prohibit[ed] the city from taking 'concerted action to drive homeless individuals from Santa Ana' and barred officials from marking the bodies of people charged with minor offenses for identification" (Johnson 1990, B12). Simultaneously, the deputy public defender representing the arrested homeless persons refused a city plea bargain that would have allowed them to plead guilty and pay fines or serve up to a day in jail. The homeless defendants argued that their right to a trial by jury—a trial in which the very legitimacy of the sweeps and deportations could be questioned—was too important to be bargained away (Johnson 1990).

In February 1991, a municipal court judge threw out charges against 22 of the arrested, finding that the Santa Ana police "deliberately and intentionally implemented a program which targeted those persons living in the Civic Center, the homeless" (quoted in Eng 1991 A1). Even though it was only homeless people who were rounded up and charged with crimes such as littering (in the case of one, for dropping a cigarette butt on the ground), Chief Walters again insisted that his department was not targeting the homeless but only those who violated the law (Eng 1991). A deputy district attorney reiterated that such arrests were vital, since the city was insistent on applying "the broken window theory" (Eng 1991, A1). The judge in the case dismissed such reasoning: "If the Civic Center is to be compared to a house and the broken windows to minor offenses, all broken windows must be repaired. In this house, it is apparent that only those consisting of plain pane glass were handled, while those of bevel or stain glass were not" (quoted in Eng 1991, A1). As the judge made plain, the arrests were conducted for the sole reason of "harass[ing] and intimidat[ing]" the homeless. Eventually the homeless won a $400,000 settlement from the city.

While the chief of police asserted that his department was not targeting the homeless, city officials continued to search for ways to push homeless people out of the Civic Center (Takahashi 1998, 157–165). The tool they turned to was an anti-camping ordinance, which they hoped might prove both more effective and cheaper than the previous, more heavy-handed, approach had proved.[14] The first anti-camping ordinance, passed in August 1992, made it illegal to sleep in a sleeping bag or blanket or to store belongings on public property, and specified a sentence of up to 6 months in jail for violators. Apparently the ordinance also asserted that "homeless people were to clear out of town by sunset" (Di Rado 1994, A27, quoted in Takahashi 1996, 158). Eleven months later, the 4th District Court of Appeals in Santa Ana temporarily enjoined the city from enforcing the anti-camping ordinance, writing that it was too "vague, impose[d] class-based restrictions on the ability to live and travel, and constitutes cruel and unusual punishment" (Di Rado 1994, A27, quoted in Takahashi 1998, 159).[15]

The city responded 2 weeks later by reviving the 1872 California vagrancy law, but was soon restrained from enforcing this as well.[16] Finally, the city wrote an anti-camping ordinance based on a U.S. Park Service ordinance that had already been upheld by the U.S. Supreme Court. One of the key aspects of this law was that it would govern only the Civic Center and not the city as a whole (Takahashi 1998, 160), a geographical restriction that is crucial to the law's validity.

Anti-camping ordinances have become a favored tool of cities seeking to remove homeless people from public spaces. These laws typically make it illegal to sleep or "lodge" in any public place, either within a jurisdiction as whole or, as with the third Santa Ana law, in a specified area. Many have specific exemptions for occasional dozing. Santa Ana's law referred specifically to camping, rather than sleeping, making it "unlawful for any person to camp, occupy camp facilities, or use camp paraphernalia" in public spaces.[17] The only anti-camping ordinance to be examined by the Supreme Court was that of the National Park Service (NPS). The NPS had outlawed camping in a number of Washington, DC, parks, including Lafayette Park across the street from the White House. In the early 1980s the Community for Creative Non-Violence (CCNV), a DC-based homeless advocacy organization, defied the ban as a specific political protest against Reagan administration policies that its members perceived to be leading to the rise of homelessness. After homeless protesters were arrested, CCNV contested the anti-camping ordinance on First Amendment grounds, claiming that their Lafayette Park encampment was a form of speech that should be protected. The Supreme Court did not deny that the encampment and the conduct of the homeless people had "expressive content" and thus was subject to increased First Amendment scrutiny. Nonetheless, the Court upheld the anti-camping ordinance, in part because camping was permitted in other parks, thus allowing homeless people to legally make their protest there.[18] Restricting camping in Lafayette Park was, in the eyes of the Court, a reasonable "time, place, and manner" restriction on public protest, and it was one that serves a legitimate government interest (the maintenance of the park for other uses) (*Clark v Community for Creative Non-Violence* 1984).[19] Santa Ana took its cue from this case and wrote its law so that it only covered a portion of the city, presumably leaving open outlying public areas for sleeping.

Even so, the law was promulgated within a quite complex and confused judicial milieu.[20] Around the time the Santa Ana City Council was crafting its third anti-homeless law, a federal court in Miami handed down a decision in a case similar to those already settled by Santa Ana. In Miami a group of homeless people contested the city's policy of arresting (or simply harassing) homeless people for engaging in what the court called "essential, life-sustaining acts," including sleeping in public, standing around, and eating (*Pottinger v. City of Miami* 1994, 1554). The Court found that, since the number of shelter beds was not even close to sufficient for the number of homeless people in the city, homeless people "simply have no place to go" but public space (*Pottinger* 1994, 1554). Therefore, the Court reasoned, arresting or harassing homeless people for engaging in necessary acts, which necessarily had to occur in public, constituted punishment for a *status* (the status of homelessness) and thus constituted cruel and unusual punishment as it had been defined in a celebrated 1962 Supreme Court decision (*Robinson v. California* 1962). As a result, the Court ordered the city of Miami to create two "safe havens" in which the city was enjoined from "arresting homeless individuals who are forced to live in public for performing innocent, harmless, inoffensive acts such as sleeping, eating, lying down or sitting . . . " (*Pottinger* 1994, 1584).

Meanwhile two other cases were working their way through federal courts, each of which was decided in 1994. In Dallas, a federal district court once again found that the number of shelter beds was inadequate to the demand and hence that a prohibition against sleeping in public punished a status rather than an act. Indeed, the court held that the Dallas anti-camping ordinance punished not just the status of homelessness but also the "status of being": "Because being does not exist without sleeping, criminalizing the latter necessarily punishes the homeless because of their status as homeless, a status forcing them to be in public" (*Johnson v. City of Dallas* 1994, 350).[21] By contrast, in San Francisco, a federal district court held that anti-homeless laws did *not* punish status. This case tested the validity of San Francisco's Matrix program and, so, touched on a whole suite of anti-homeless laws and police practices. The judge in the San Francisco case held that homelessness was not a status (and so laws regulating the behavior of homeless people were not cruel and unusual punishment) because homelessness was not an "immutable" condition. The judge held that the presence of the characteristics that defined the status under question should be present "at birth" (like gender or race) or be truly out of the control of the individual (*Joyce v. San Francisco* 1994). In this latter category were such things as illness and drug addiction (following the *Robinson* decision). While drug addiction might seem an odd characteristic to accord "status" (had the Supreme Court not already done so), and an even harder characteristic to find substantially different from homelessness, the judge held that it was indeed different from homelessness since a homeless person "immediately loses her 'status' when provided housing" (see J. Smith 1996, 327) while an addict always remains an addict (though perhaps a recovering one). The presumed "status" of homelessness, therefore, was neither immutable nor long-lasting.

Into this milieu the Legal Aid Society and the ACLU launched their case against the Santa Ana ordinance. In the first instance the homeless and their advocates won. Following the Robinson Doctrine, the California Court of Appeals restrained the city from "enforcement of the ordinance in its entirety" (*Tobe v. Santa Ana* 1994, 387; J. Smith 1996, 324). The city appealed to the California Supreme Court, which reversed the appeals court decision. But it did so not on the Eighth Amendment grounds of the Robinson Doctrine but rather because it found that lawyers for the homeless had "failed to perfect an 'as applied' challenge" (*Tobe v. Santa Ana* 1995, 1150; J. Smith 1996, 324). The court therefore tested the constitutionality of the ordinance "on its face," which means that it tried to determine whether the ordinance was constitutional in and of itself rather than in how it was as applied to a specific class of people—the homeless (J. Smith 1996, 324).

In making its decision, however, the California Supreme Court left open the possibility of a new kind of defense of the "rights" of homeless people. The court suggested that a "defense of necessity" might be raised if for some of the homeless "violation of the ordinance is involuntary" (*Tobe v. Santa Ana* 1995, 1155; Fasanelli 2000, 337). Against "persons who . . . have no alternative to 'camping' or placing 'camp paraphernalia' on public property," the Court argued, the ordinance should not be enforced (*Tobe v. Santa Ana* 1995, 1155).

In one of the original cases that were brought together as *Tobe v. Santa Ana*, James Eichorn was charged in January 1994 with violating Santa Ana's anti-camping ordinance. Unlike the rest of his codefendants, Eichorn insisted that his case go to trial. The trial court ruled that the necessity defense was unavailable because Eichorn had not shown that he broke the law only to "avoid a 'significant, imminent evil' " (Fasanelli 2000, 345, quoting In re Eichorn 1998, 536). In response to the Tobe decision, a California Court of Appeals found that Eichorn should indeed have been allowed to assert the necessity defense in his original trial, arguing among other things that "by sleeping in the civic center, the defendant may have been avoiding the 'significant evil' of sleep deprivation" (Fasanelli 2000, 346, quoting In re Eichorn 1998, 539). But this finding in turn rested on the fact of a significant shortage of shelter or other housing in Santa Ana. In 1993 the city provided only 118 permanent shelter beds plus an additional 125 during the winter— all for a population of perhaps 1,500 homeless people in the city (Fasanelli 2000, 347).[22] On the night that Eichorn was arrested, the armory, where the winter beds were located, was filled beyond its capacity. The appeals court in In re Eichorn, drawing on facts such as these, made it clear that necessity is a reasonable justification for violating "quality of life" laws.

As Antonia Fasanelli (2000, 350) concludes, "In other cities, where courts have upheld anti-camping and sleeping ordinances as constitutional, the necessity defense will [now] be available to homeless people under the Tobe/In re Eichorn theory if the violator shows that more homeless people than shelter spaces exist and there is a lack of adequate income to pay for housing." If the cities surveyed by the National Law Center on Homelessness and Poverty (NLCHP 1995; 1997) are even close to being representative, then there is not a medium- or large-sized city in the United States where this is not the case.[23] "Fixing Broken Windows" in the way that Kelling and Coles advise, then, can be seen even more clearly as a punitive policy directed against a despised class rather than anything like a reasonable urban policy.

ANTI-HOMELESS CAMPAIGNS AND THE CONTENT OF CONTEMPORARY URBAN JUSTICE

There is something exceedingly perverse in the above discussion, something easily lost as the details of specific cases are outlined—namely, that homeless people and their advocates are driven, in the current urban context, to argue for the right to sleep in public, to lie on sidewalks, to beg on the streets, or to shit in alleys. These are pretty mean, pretty shriveled, rights. As an attorney for homeless people in New York City remarked after winning an early decision (later overturned) that secured the right of beggars to panhandle in New York City subways and stations: "It's hard to get real excited about winning the right to beg" (*New York Times* 1990, B1). In that light, take a moment to consider just what the California Supreme Court achieved as it adjudicated a concerted campaign by the city of Santa Ana to rid itself of homeless people: it opened the door for people to show that, in the normal course of events, they have no choice but to break laws that most of us would find perfectly reasonable (such as not camping in a civic center). The normal state of affairs in the American city is such that one can *raise the question* as to whether some number of residents have no choice but to live in the open, to satisfy all their bodily needs in public, to go about begging. It has forced us to map the city to see if there are some few areas in which people have the right not only to sleep or eat, but just to *live*, and to live a life defined by physical and mental illness and the constant threat of death by exposure or at the hands of thugs. It has asked us to scour the city to see just how close we are to the "libertarian paradise" that Waldron (1991) described, the paradise in which every space is governed by something like a private property rule such that there simply are no public spaces in which those without access to private property, or its corollary, highly policed public space, simply can be.

Some judges—the California Supreme Court justices who left open the door of necessity among them—have recognized, at least to some extent, that such a "paradise" is flawed at the most basic level of human rights. Traveling the farthest along this road was the federal district court in Miami (cited above) that ruled that the city had to establish two "safe havens" on public property where homeless people could stay free from the fear of harassment by the police. As with the California court, this ruling produces a pretty mean notion of urban justice, one that does not even contemplate ordering a decent housing program. Yet, even so, it nonetheless announces the limits of the broken windows thesis and public space policing more generally by creating a ghetto for homeless people in which at least a few shreds of their rights will be maintained, even if, within the larger context of the city as a whole, those rights are pretty much abandoned.

The Miami court's order to create safe havens, like the California court's recognition of the necessity defense, in essence proposes a geographical solution for what is a social—and political-economic—problem. In the case of Santa Ana, which carefully wrote its anti-camping ban so as to cover only a portion of the city, it will not be hard to argue that, in fact, Eichorn and others could have slept somewhere else, somewhere outside the civic center, and hence the necessity of sleeping in the civic center itself cannot be shown. Picking up on the geographical logic of recent court cases, Yale legal scholar Robert Ellickson (1996) has influentially argued that, rather than outlawing unwanted behaviors altogether, cities should instead develop a process of "public space zoning." It is worth looking at Ellickson's argument in some detail for two reasons. First, it offers compelling insight into how regulating space—creating "proper" urban geographies—can be easily substituted for concerted progressive policies designed to attack *social* problems and to expand the content of urban social justice. Second, it provides insight into how "broken windows"-style regulation of urban space is likely to develop in the near future. It provides insight into the new regime of order—and hence the new regime of rights—that is likely to mark the American city of the early 21st century.[24]

PUBLIC SPACE ZONING

Robert Ellickson is perhaps most noteworthy for his book *Order Without Law: How Neighbors Settle Disputes* (1991), in which he makes the argument that "informal" controls on social order are often more effective than legal ones. Ellickson is leery of the state, particularly at the national scale but to some extent at the more local scale, as well. Property owners, he asserts, have a greater stake in preserving qualities and values than do state bureaucracies. Given this perspective, he holds that the management of land needs to vary spatially in accordance with neighborhood, city, and state "norms," and that it should not be subject to universalizing federal constitutional oversight. In 1996, relying explicitly on the "broken windows" thesis, Ellickson extended his argument to the management of urban public space.

In what has since become an influential and heavily debated article, "Controlling Chronic Misconduct in City Spaces: Of Panhandlers, Skid Rows, and Public Space Zoning" (1996),[25] Ellickson adheres to Kelling and Coles's (1996) injunction that questions of order in public space must be disconnected from issues of homelessness—by sheer force of will if in no other way. Hence, though he contradicts himself in several places in the article (by pointing out, for example, that street people are in fact destitute and quite often homeless), Ellickson argues that "homelessness" is "an unduly ambiguous word" that "implies policy solutions that are inapt" (1996, 1192) and that "tends to entrap [homeless people] in a marginal status" (1193). "Homelessness" is the wrong descriptor because what is at stake, he argues, is not poverty or housing, but what he calls "chronic misbehavior." "Chronic misbehavior" on city streets needs to be understood as the product of two types of people: those Ellickson calls "bench squatters" (those who "monopolize" park or other benches and sidewalks with their belongings and bodies, be they homeless "bag ladies" or "Proust readers" [1184]); and those he calls "chronic panhandlers" (those who beg in the same place day after day). Ellickson asserts that his policy proposals target these forms of *street conduct* and do not address the status of homelessness.

To the argument that anti-homeless laws target activities that homeless people have no choice but to engage in—that is, they target the involuntary *status* of being homeless rather than the specific acts they purport to regulate—Ellickson responds that "to treat the destitute as choiceless underestimates their capacities and, by failing to regard them as ordinary people, risks denying them full humanity" (1187). Begging, therefore, needs to be understood as "an option, not an inevitability" (1187). Indeed, according to Ellickson, beggars and bench squatters are *more free* than the rest of us because, not "living lives structured around families and employers," street people have more time to "individually craft a daily routine" (1187) and "move from place to place" (1188).[26] Ellickson even argues that ordering a mentally ill woman "squatting" on a bench to "move along" "might actually enhance the liberties of the mentally ill" because "she herself might prefer that outcome to bearing the risks of involuntary confinement" (1189) in jail or in an institution.[27] This argument is best read perhaps in light of an earlier article by Ellickson (1990) in which he argued that providing decent shelter to homeless people actually causes an increase in homelessness.

Regulating Public Space: Norms and Harms

We have already seen that proponents of the "broken windows" thesis have had a difficult time deciding whether "chronic misbehavior" actually rises to the level of an identifiable "harm." Ellickson avows that it does but that the *degree* of harm may vary across urban space. He asserts, therefore, that cities need to establish a program of public space regulation that does not rely on the universalizing tendencies of either law or rights, arguing that these "succumb to the notion that all open access spaces have to be governed by an identical regulatory regime" (1996, 1219, n. 301) and thus are inadequate to the differing needs of communities. Most cities' regulations are spatially insensitive: "A constitutional doctrine that compels a monolithic law of public spaces," Ellickson writes, "is as silly as one that would compel a monolithic speed limit for all streets" (1247). Instead, he suggests that a spatially variable regime of urban public space zoning needs to be developed. A "city's code of conduct," he argues, "should be allowed to vary spatially—from street to street, from park to park, from sidewalk to sidewalk" (1171–1172). Optimally, this zoning should be "informal," that is, developed by the "community" as it establishes "norms" of behavior for people who use public space in its midst (1222–1223). This informal zoning should then be maintained by "trustworthy police officers" (1173, 1245) who enforce the norms the community has established. The second-best solution is for cities to create a formal system of public space zoning that allows for different sorts of behavior—and perhaps even some degree of "misconduct"—in the various public spaces of the city (1246).

Ellickson's target, the thing that needs policing, is "chronic street nuisance," which he defines as "behavior that i) violates community norms governing conduct in a particular public space ii) over a protracted period of time iii) to the minor annoyance of passers-by" (1175). That is, he is interested in regulating behavior that even does no more than create a sense of "minor annoyance"—perhaps just a cracked, rather than a broken, window. But for Ellickson, such cracked windows—such bench squatting and chronic panhandling—represent a set of real "harms" (1177) to the general community. First, and circularly, annoying street behaviors may "trigger broken windows syndrome . . . signal[ing] a lack of social control" (1177). Second, and more specifically, Ellickson notes that since many public authorities have taken to eliminating or redesigning public benches to discourage "bench squatting," a "proliferation of bench squatters . . . tends to lead to the elimination of amply sized benches" from public space (1178, n. 50). (The direction of causality is not just insulting; it is symptomatic of a whole mode of reasoning concerning the homeless.) Third, Ellickson suggests that pan-handling "worsen[s] race relations in cities where panhandlers are disproportionally black" (1181). (He also makes the appalling argument that one of the primary results of the civil rights movement was to make it easier for African Americans to live as homeless people on downtown streets.[28]) And fourth, according to Ellickson, "begging signals an erosion of the work ethic," a "harm" that "all human societies" attempt to remedy (1182).

Given these "harms," Ellickson proposes the following rule: that "a person perpetrates a chronic street nuisance by persistently acting in a public space in a manner that violates prevailing community standards of behavior to the significant cumulative annoyance of persons of ordinary sensibilities who use the same spaces" (1185). How should "community standards" be determined? Here Ellickson turns to Jane Jacobs: "The first thing to understand is that the public peace—the sidewalk and street peace," Ellickson quotes Jacobs (1961, 31–32) as saying, "is not primarily kept by the police, necessary as the police are. It is kept primarily as an intricate, almost unconscious network of voluntary controls and standards among the people themselves and enforced by the people themselves" (Ellickson 1996, 1196). Jacobs's argument serves Ellickson well because he strives to show that what needs to be instituted are putative *community* norms. Yet since, as we will see, Ellickson has a remarkably truncated notion of who belongs in a community, he quickly discards Jacobs's argument in favor of promoting the police themselves as the primary guarantors of public order (1173, 1200–1201, 1208–1209, 1245). But his point in invoking Jacobs is not at all to debate the merits of city policing; rather, it is to deflect attention from that issue and to instead invoke a nostalgic vision of the city that serves as the template for the sort of public space zoning he wants to promote.

Skid Row: Ellickson's Nostalgic City

This nostalgic vision is of a time when almost all American cities had within them what Ellickson calls "informally policed Skid Rows" (1996, 1208): the 1950s. In Ellickson's view skid row in the 1950s was a place "along with closely related Red Light Districts . . . where a city relaxed its ordinary standards of street civility" (1208). This is a quite partial view of skid row—the bulk of the evidence suggests that it was always a heavily and stringently policed place (Anderson 1923; Bahr 1970, 1973; Bittner 1967; Blumberg et al. 1978; Foote 1956; McSheehy 1979; Wallace 1965; J. Wilson 1968)—but it does allow Ellickson (1996) to make a curious, if wholly unsupportable, point that will become central to his whole argument—namely, that skid row not only was the appropriate home for alcoholics and the elderly poor, but that it made it possible for the police to act benevolently in their guaranteeing a *diversity* of social orders (1172, 1202–1209). Here is what Ellickson says:

> In Skid Row . . . moderate public drunkenness was likely to be tolerated, not only by the other down-and-out residents, but also by the police.[29] By contrast, the same level of inebriation elsewhere in downtown was much more likely to get an alcoholic in trouble. In the 1950s, a cop on the beat might unhesitatingly tell a "bum" panhandling or bench squatting in the central business district to "move along."[30] A bum on a Skid Row sidewalk would never hear this message because he was exactly where the cop wanted him.[31] In this way, the 1950s police officer helped to informally zone street disorder into particular districts. (1208–1209).

Ellickson's own sources directly contradict him, showing how the police did in fact make frequent arrests on skid row, and told the men and women there to "move along" (Foote 1956; Schneider 1986; J. Wilson 1968). And other sources, taking the ethnography of skid row into the 1970s (a period in which, Ellickson avers, policing of skid row was unduly hampered by constitutional restrictions on police power), show that the police could be impressively brutal in their use of arrest as a disciplinary mechanism on skid row (McSheehy 1979). But, never mind.

For Ellickson (1996), it is not the brutal policing of homeless men that was a problem; he simply dismisses that on the grounds that it was less brutal on skid row than in other parts of the city (1208–1209, n. 232–234). Rather, the "constitutional revolution" (1209) of the 1960s and 1970s—that period in which vagrancy laws were found unconstitutionally vague and status crimes were decriminalized—had the effect of "nationalizing" laws concerning street disorder (1209). That is, judicial liberalization and the recourse to constitutional law to litigate arrests for public drunkenness, vagrancy, loitering, and the like applied a single standard of justice across all the urban spaces of the country. Such a "nationalization," in Ellickson's estimation, created a system that was "centralized and inflexible" (1213), making the sort of "informal zoning" that he thinks marked the 1950s skid row impossible (which, of course, was precisely the point). Thus, and also because so many churches gravitated to the suburbs (1216), skid row fell into decline,[32] and the visible evidence of the "down-and-out" life diffused across the other spaces of downtown: "Street people who had previously been informally confined to Skid Row were now able to make chronic use of the busiest downtown areas. Many of them did" (1216).

Zoning Public Space

Ellickson's (1996) nostalgia leads him to an intriguing proposal, one that, taken at face value, seems to argue for a return of skid row (and homeless people's sequestration there). However, as interesting and important as it is, it is a proposal that simply should not be taken at face value, since it is, as we will see, thoroughly disingenuous. Ellickson proposes that cities should zone their public spaces into three categories: red, yellow, and green (1220–1222). In red zones, which he argues should constitute perhaps 5% of a downtown area, "normal standards for street conduct would be significantly relaxed" (1221). "In these relatively rowdy areas," Eillickson (1221) writes, "a city might decide to tolerate more noise, public drunkenness, soliciting by prostitutes, and so forth." Red zones would serve as "safe harbors for people prone to engage in disorderly conduct" (1221). Yellow zones, covering about 90% of downtown, should "serve as a lively mixing bowl" (1221). Here, "the flamboyant and the eccentric" (1221) would be allowed in, just so long as they did not overstay their welcome. Here too "*chronic* (but not episodic) panhandling and bench squatting . . . would be prohibited" (1221). Finally, green zones, occupying the final 5% of downtown space, would become "places of refuge for the unusually sensitive: the frail elderly, parents with toddlers, unaccompanied grade-school children, bench-sitters reading poetry" (but presumably not Proust) (1221). Even *episodic* panhandling and bench squatting would be outlawed. In essence, then, Eillickson proposes to codify space such that at the scale of the city, a mix of "land uses" would be tolerated, and at the scale of the downtown 90% of the area would serve as a "lively mixing bowl" of peoples and activities, all overseen by a benevolent police force working to maintain the "community norms" of each area.

"Community norms" are the key:[33] the determination of red, yellow, and green zones, according to Ellickson, should be based on what a community wants. Zoning of public space should be done informally by "members of a close-knit group who repeatedly make use of open access public space" (1222). These members should "enforc[e] social norms to deter an entrant from using [public space] in a way that would unduly interfere with the opportunities of other members" (1222). Such community members, according to Ellickson, are particularly adept at "recogniz[ing] the crazy-quilt physical character of urban spaces and the myriad demands of pedestrians [and they] tend to vary their informal norms from public space to public space" (1222). In this effort, they are aided by the police, who work to enforce these varying community norms (1223). And over time, Ellickson hopes, residents and homeless people alike will internalize these rules, and the norms of the different zones will become second nature (1225–1226). Yet, even so, Ellickson recognizes that informal zoning might not be effective. Though he does not say it outright, it is clear that Ellickson is concerned that "chronic panhandlers" and "bench squatters" could be taken for *members* of the "close-knit group"—those who repeatedly make use of public space—and so he suggests that city governments should be given leeway to formally zone public space into red, yellow, and green zones, complete with signs listing applicable rules.

Despite this worry about the efficacy of informal zoning, Ellickson still puts his faith in "the community" for whom public space will be policed. Yet—and this is crucial—Ellickson *never* explicitly defines community. What emerges, in the course of his long law review article, however, is that this "community" simply does not include homeless people. They are in no sense considered to have any a priori rightful claim to the use of streets and parks: they are figured only as unwanted nuisances. They have no standing whatsoever as members of the communities in which they live. So who then is included in this community? In the only hint at the community he has in mind, Ellickson points to various "individual champions of the public": pedestrians, owners and occupiers of abutting land, and organizations that enforce street decorum (such as Business Improvement Districts and the police) (1196–1199). With the exception of "orderly" pedestrians (1197) and the police who work in the interest of the "community" as a whole, Ellickson's community is apparently the community of *property*, since, as he shows, it is property owners (and to some extent those others such as renters who are covered by private property rules) who suffer the greatest "harm" from nearby homeless people engaged in little more than "minor annoyances."

Yet Ellickson must recognize (since much of his research has been on issues of zoning and land management) that this community of property (including renters) will fight against red-zone designation: no community of property would willingly accept such a status, as it would incur unacceptable costs in terms of falling property values and increased maintenance and service costs.[34] The "negative externalities" attendant upon the creation of an official ghetto (with its designating signs) would be too great. His advocacy of "informal" zoning, therefore, stands as all the more curious and fanciful: he never addresses the questions of why informal zoning would not suffer the same problems as formal zoning. Why would property owners (or adjacent property owners at the edges of the district) not resist the decline of property values attendant upon the harboring of "broken windows"? The answer is that they would not,[35] and thus the development of freely tolerated—not judicially mandated—"red zones" is extremely unlikely, especially since, in Ellickson's view, "the first best solution to the problem of street misconduct would be the maintenance of a trustworthy police department whose officers would be given significant discretion in enforcing *general standards* against disorderly conduct and public nuisances" (1245, emphasis added). Ellickson never even broaches the question of how "general standards" and "community norms" are to be determined. In Ellickson's proposal there is simply no mechanism—and certainly no *democratic* mechanism—outlined for instituting informal zoning, much less for guaranteeing any level of spatial justice. Not coincidentally, then, Ellickson suggests absolutely no means—legal, constitutional, legislative—for guaranteeing that *any* space would be zoned red.

And, one can surmise, this is precisely *why* Ellickson so hopes to win approval for his plan of informal zoning. It is, in the end, an elaborate hoax, but an extremely dangerous one. It leads exactly to the same hoped-for outcome as anti-homeless laws in general: the elimination of not homelessness, but homeless people, by eliminating any space in which they can *be*. The differentiation of space, this practical plan for the implementing a "lively mixing bowl" in the city, *is* nothing more than the desire for the same purified space that ultimately motivates anti-homeless laws. But with this crucial difference: to the degree that Ellickson's plans for informal zoning are adopted, they will remain out of the purview of the courts, eliminating a crucial arena in which homeless people's right to the city can be struggled for, even if those same courts are sometimes inhospitable to such claims of right.

CONCLUSION

In fact, we have already seen how well the zoning of public space along the lines of Ellickson's suggestions works in the contemporary city. Instead of merely appreciating or criticizing Ellickson's public zoning proposal as a "thought experiment" (as he calls it in one place), we would do well to return to the history of People's Park and Telegraph Avenue in Berkeley. There we can see just how "community norms" are constructed: they do not just arise spontaneously; if they exist at all, they are the result of serious and concerted social struggle. Actors with differing degrees of power, including large institutional players such as universities, city governments, merchants with differing views on what makes a "lively mixing bowl," activists, and the homeless themselves all contend over just what the "norms" of the community shall be. The police too are involved in this process, not just as enforcers of already established community norms but sometimes also as their progenitors or their transgressors. Commenting on it all are any number of "little Arnolds" who seek to sway public opinion—to influence social norms—toward a more, rather than less, restrictive ordering of public space. The concept of social norms, in other words, misses exactly the "dialectic of public space" (as I called it in Chapter 4) that develops through struggle over particular places and the implementation of particular social visions. Ellickson's view of the ordering of public space simply ignores the forces at work on the ground in all public spaces, forces that are as contentious as they are consensual. And the Berkeley case shows just how important an appreciation for *power* must be when considering how public spaces are to be ordered and policed: we must always be aware of who benefits from social order and consensus and who doesn't, whose interests are served and whose are not.

We have also seen how zoning works in the context of the two abortion protest cases and the history of public forum jurisprudence examined in Chapter 2. There we saw how the formal regulation of public space was both a means of institutionalizing the rights to free speech, assembly, and protest and a means of undermining exactly those rights. Most importantly, we saw that this dialectic only developed and progressed to the degree that people violated established laws, laws that most frequently sought to protect state and corporate power in the name of upholding speech and assembly rights. Ellickson's call for informal or formal public space zoning simply ignores the social history

that is at the root of the constitutional "revolution" Ellickson so de-
plores. Rights to public space, as we have seen throughout this volume,
have only been expanded when they have been forcefully demanded,
quite often by people breaking the existing laws and thereby showing
those laws—about picketing as much as about sleeping—to be oppressive, in their geography if not in their actual wording. The IWW's demand for the right to speak on the streets of San Diego, Fresno, and
other cities of the United States was, obviously, in direct contradiction
of "community norms" (Chapter 2). Indeed, that was exactly the point.
And through that contradiction, the IWW was able to show just what
was wrong with those norms (at the same time its members struggled to
transform the larger political and economic context that allowed them
to be norms in the first place). And, as I have shown elsewhere (in the
case of anti-picketing laws in agricultural California in the 1930s:
Mitchell 1998c), the very construction of "community" is itself a political project often accompanied by great violence. Nor does the abstraction of "community norms" provide any purchase on the difficult questions attending the zoning of public space outside abortion clinics.
Indeed, in such places it is hard to imagine what *community* norms
could possibly be since, in Ellickson's telling, the community he is interested in is a community of consensus (overseen by property). Such a
community in the case of abortion and abortion providers, for example,
will necessarily run roughshod over *someone's* rights—and not through
any democratic means, either.

But in conclusion to this chapter I want to return to the question
that developed as the heart of the Santa Ana cases and which Ellickson
is so happy to ignore: the question of necessity and its implications for
the right to the city. If we recall one of Lefebvre's arguments outlined in
Chapter 1, we remember that any reasonable "right to the city" requires
also a right to *inhabit* the city, a right to housing. For Lefebvre (1996
[1968], 179), the right to housing was a necessary precondition of the
right to the city. So too has it long been for activists. But what is remarkable about the contemporary city, the city that allows a proposal like
Ellickson's to pass as a reasonable policy suggestion, is that the right to
housing simply is no longer even considered legitimate. In this regard,
consider again the decision the California Supreme Court made in *Tobe*.
In a part of the decision the court recognized that there is indeed a fundamental right to travel implicit in the U.S. Constitution and that this
right to travel carries with it a concomitant right to stay put, a right to

remain in place. But the court also argued that the existence of such a right in no way obligates the government to find a means to assure that people can exercise that right. Neither the city nor any other jurisdiction had to provide the homeless with a place to stay. Rather, it was the obligation of homeless people to legally secure such a place. In the abstract this might make some sense; in any actually existing housing market, however, it is sheer nonsense.

Yet it is also quite typical of current reasoning, quite typical of a world in which the right to shelter—or better, the right to housing—is no longer considered an arena for state intervention. The always minimal U.S. commitment to housing the poor, enshrined first in the public housing act of 1937, has been all but gutted. This is particularly interesting, given just how easy it has become for homeless people such as James Eichorn (see pages 208–209) to show that they necessarily have to break laws in order to live.

It is also indicative of a larger transformation of the economics and politics of the contemporary city. If one of the solutions to the crises of overaccumulation that marked the Great Depression was to implement new systems of collective consumption—such as subsidized housing—both to jumpstart certain sectors of the economy (construction, consumer durables) and to effectively drive down the costs of labor to individual capitals (by subsidizing the real cost of labor reproduction), then by the late 1960s this solution was increasingly understood to itself be a fetter on continued capitalist expansion and accumulation (Harvey 1982, 31). As productive capital was "globalized" and labor markets internationalized (such that the reproduction costs of labor were often borne elsewhere [N. Smith 2000]), welfare and housing subsidies in the United States lost both political favor and their political–economic raison-d'être. In this light, the famous implosion of the Pruitt-Igoe public housing project in St. Louis in July 1972[36] represented not only the end of an architectural era (Jencks 1981; Harvey 1989) but also the inauguration of what Harvey (1982, 31) calls a new "class strategy." This strategy has sought to capture relative surplus value through the gutting rather than the development of social services and other forms of working-class subsidization, in the expectation that these costs will be depressed through the simpler mechanism of immiseration. Under this new neoliberal political-economic model, the social costs of labor reproduction are, like the benefits of development, privatized.

The contradictions that such a shift in strategy has led to are apparent: the rise of homelessness and other markers of abject poverty are, of course, a means of instilling discipline in working populations (Piven and Cloward 1992; Peck 1996), even as they also seem to threaten a city's ability to capture investment in an increasingly competitive global market in corporate locations. The response to this contradiction can take two forms, one of which, as we have seen, has been fully exercised: this is the strategy of criminalization of poor and homeless people, of greater regulation of space, and of a minimalist discourse about rights (arising in response) that is restricted to discussions over whether or not people should have the right to urinate in alleys, to sit on sidewalks, or to sleep in public parks. The second strategy is one of struggling for a greater right to the city, a right that includes the right to housing, the right to space, and the right to control, rather than be the victims of, economic policy. This second strategy requires, however, a reclamation of public space, not for societal order and control (as important as those, in fact, may be) but rather for the struggle for justice. It requires, as we saw in our discussion of Iris Marion Young's ideas in Chapter 1, a strong commitment to distributive justice, but not only that. It also requires taking control over the *means* of distribution—and production—of justice. In turn, this requires that we return again to Hyde Park, that we find ways to turn the space of the city into the site of the demand for justice—a justice that *requires* housing as a precondition but that also requires even more: the incessant cry and demand for the right to the city.

NOTES

1. The number of annual deaths in any city is highly variable. In 1997 San Francisco congratulated itself for having witnessed what the *San Francisco Chronicle* (1997) called a "big drop in homeless deaths." Homeless deaths "plunged 34 percent" from 1996 levels. Still, 102 homeless people died on the streets: nearly two a week. The year 1993 likewise saw a drop in homeless deaths, with 101 dying. In 1992, 138 died on the streets (*San Francisco Chronicle* 1993). San Francisco tracks its homeless deaths from December 1 to November 30 each year.
2. *Tobe v. Santa Ana* (1995), 1177. The Santa Ana case has been widely analyzed. Among others, see Takahashi (1998, Ch. 8); Simon (1995); Fasanelli (2000); J. Smith (1996). We will come back to the details of this case throughout this chapter. The history of demonization of "vagrants" in the United States is traced in Cresswell (2001).

3. Such a result, of course, is not unique to homeless people. The controversy over racial profiling that commanded much press attention before the September 11, 2001, terrorist attacks, no less than the almost immediate reconsideration of whether profiling was so bad in the wake of the attacks, is indicative of how fragile human rights can be for "suspect classes." The Bush administration, and especially the Department of Justice, seems intent in its "war on terrorism" to make clear just how conditional rights can be when a state is not forced to protect them. The range of detentions, deportations, and denials of basic rights the federal government oversaw in the months after September 11, especially as a matter of policy, is only now becoming clear. The willingness of the courts to support even the most absurd restrictions on immigrants, radicals, and others was made plain enough during World War I (see Chapter 2) and since then with the "relocation" of Japanese Americans during World War II; COINTELPRO (the Counter Intelligence Programs of the FBI) and other actions against radicals during the 1960s and the abrogation of search-and-seizure protections as part of the "war on drugs" all give little reason for optimism that either the federal state or the courts will find it within their power to keep from instituting the most draconian "security" measures—unless, of course, forced to do so.

4. And so here the architects of the "broken windows" thesis reveal themselves to be as concerned with creating a city filled with conventional, acceptable people as with reducing crime. They are concerned with eliminating people whom they see as "out of place" in the terms established by Cresswell (1996).

5. Kelling and Coles (1996) specifically argue that engaging in indiscriminant street sweeps is bad public policy (not because it is wrong but because it leads to bad publicity), but this has not deterred those, such as Santa Ana's Police Chief Walters, who (as we will see) used the broken windows thesis to justify such sweeps.

6. This is, of course, the logic of the ghetto, a logic that the Nazis pushed to an extreme, but which they were by no means alone in adhering to (as the internment of Japanese Americans during World War II made plain).

7. That is, space as *landscape* (as discussed in Chapter 5) is determinant.

8. Beginning in the late 1950s, shifts in American constitutional law led to the repeal of most status crime laws such as vagrancy laws that punished people for an involuntary status rather than for some identifiable conduct. By clearly defining "disorderly behavior" as those actions that homeless people must engage in (like sleeping or sitting on benches and sidewalks), authors such as Wilson and Kelling are calling for the return of status crime punishment under a different guise. The argument about harms that public order proponents present will be dissected more fully below.

9. We will examine these claims more fully below.

10. The argument here, of course, is circular, since Kelling and Coles (1996) are making it in defense of instituting laws that make many of the behaviors they dislike illegal.

11. Alice Baum and Donald Burnes are the authors of an analysis of homelessness, *A Nation in Denial: The Truth about Homelessness* (1993), that made quite a stir among editorialists in 1993 by blaming homeless people for the ills that befell them.

12. In fact, the evidence from the text of *Fixing Broken Windows* shows just how difficult it is to maintain the pretense that the authors are not interesting in targeting the homeless as a class. While they are often assiduous in their placing of the world "homeless" inside quotation marks to indicate that they are using the word as a shorthand for a set of behaviors, and while they occasionally profess their concern for homeless individuals and their housing needs, they also refer un-self-consciously to those they wish to see eliminated from the streets and subway stations as "indigents" (117).

13. A news article from early 1990 (Eng 1990a) cited "homeless advocates" as saying there were about 10,000 homeless people in Orange County, of whom about 1,500 were in Santa Ana. By the mid-1990s, the estimate was 12,000–15,000 homeless people in Orange County (Dolan 1995, cited in Takahashi 1998, 161). As we will see, even a range of estimates as large as this make little difference, given the documented paucity of shelter beds and the high rents in the area.

14. The avoidance of costly settlements became increasingly important as Orange County was forced into bankruptcy in the mid-1990s due to its financial mismanagement (Takahashi 1998).

15. The cruel-and-unusual-punishment argument is related to the necessity defense that we will soon examine and it will be explored in more detail then.

16. On the history of vagrancy laws in the United States, see Cresswell (2001); on their use in California, see McWilliams (1971 [1939]).

17. Santa Ana City Code § 10-402. A good overview of these laws is Foscarinis (1996).

18. The spatial reasoning of the Court here is every bit as problematic as the spatial reasoning of the University of California in 1964 when it sought to establish an alternative free speech area in a lightly traveled area of the Berkeley campus (see Chapter 3).

19. A good overview of the jurisprudence on homelessness and First Amendment issues is Millich (1994).

20. This and the next paragraph are a revision of an analysis made I first made in Mitchell (1998b).

21. The district court was later reversed on procedural grounds that did not touch on the substance of its ruling.

22. See note 13 above and its associated text.

23. On the crisis of affordable housing more generally, see HUD (1999).

24. The following section is a slight revision of an analysis I first presented in Mitchell (2001b).

25. This article has been reprinted, in condensed form, as Ch. 2 in Blomley et al.'s (2000) *Legal Geographies Reader.* While that version gives a good sense

of the flavor of Ellickson's article, as always the devil is in the details, and so searching out the full law review article is well worth it.

26. For empirical analyses of this "freedom," see, for example, Rahimian, Wolch, and Koegel (1992); Rowe and Wolch (1990); Wolch, Rahimian, and Koegel (1993).

27. The nature of this "freedom" was made plain in November 1999, when New York Mayor Giuliani ordered city police to arrest and jail any homeless person in city streets or parks who refused to move along when ordered to do so. As Sartre once commented, he was never so free than at that moment on a Paris street when a Nazi soldier held a gun to his head and told him to cross the street.

28. "The softening of white hostility towards blacks during and after the 1960s seems to have allayed the reservation many underclass blacks had previously harbored about becoming chronic users of downtown spaces. In any event, the panhandlers and street homeless who began appearing in American downtowns after 1980 were disproportionately black. No fact better demonstrates the success of the post-1960 inclusionary zeitgeist" (Ellickson 1996, 1216–1217).

29. Ellickson notes in a footnote at this point that police *did* in fact frequently arrest drunks on skid row and that they engaged in regular "sweeps" of the streets of skid row, indiscriminately detaining or arresting street people, but he dismisses this evidence with the comment that even so the police seemed to be "more permissive" on skid row than in other parts of town (1996, 1208, n. 232, citing Bittner 1967). Leaving aside the questionable logic that makes injustice OK if it is less severe in some places than others, the evidence in fact does not even support Ellickson on his most basic claims about the permissiveness of the police (see, e.g., McSheehy 1979).

30. Here Ellickson (1996) adds a footnote saying, "One can only conjecture how often night-sticks were used to enforce these orders." Actually, one could read the ethnographic and historical evidence, including Ellickson's own sources (e.g., Bahr 1973; Bittner 1967; J. Wilson 1968) (1208, n. 233). Doing so provides a clear, if not appealing, picture of the content of Ellickson's nostalgia.

31. And here Ellickson's own footnote directly contradicts the message in the body of the paper. Where in the body Ellickson says "a bum would never hear this message," in the footnote he points out just how frequently they did, but once again says this does not matter because police were "more tolerant" in skid row than elsewhere. Thus, some rather brutal policing tactics are justified because they are not as brutal as they could conceivably be (1209, n. 234). This also leaves aside the question of why "a cop" has the right to determine who is allowed where in the city—why some citizens have the right to all the city and others must be sequestered in particular districts.

32. Ellickson does not examine the processes of gentrification, assumes that urban renewal followed (rather than led) skid row decline, and is skeptical

of the role of SRO (single-room-occupancy) destruction in the growth of homelessness (1216, n. 279).

33. And here, perhaps unwittingly, Ellickson aligns his scheme with the old Federal Housing Authority program of redlining. The history of community norms-based zoning that gave rise to the FHA's racist lending practices that did so much to maintain and increase segregation in American metropolitan areas is briefly told in Jackson (1985).

34. There are exceptions to this statement, such as New Orleans's French Quarter, but it is an exception only because the "relaxed" street norms and "relative rowdiness" have been fully commodified and sold to slumming middle and upper classes, conventioneers, and the like, and are not the result of poor people seeking to meet their bodily needs. San Diego's Gaslamp district similarly tolerates a stunning amount of noise and public drunkenness in the streets and on the sidewalks from conventioneers, college students, and other bar and nightclub patrons, but is increasingly ruthless in its policies of sweeping publicly drunk homeless people out of the area. I know from experience that the late-night noise on Fifth Avenue is loud enough to make it impossible to sleep in a hotel on that street even with the windows closed and that the amount of vomit staining the early Sunday morning streets is impressive. Surely these are "minor annoyances" worth policing too—except, of course, that these annoyances are deeply profitable.

35. The best analogy is probably the history of both formal and informal zoning against adult cinemas, bookstores, and strip-clubs in most American cities—and the intense land and political disputes such zoning engenders. Consider New York Mayor Giuliani's campaign to rid Manhattan of all "adult" establishments despite formal zoning laws that allow them. Another analogy is the history of "not-in-my-backyard" (NIMBY) struggles to keep halfway houses, shelters, and the like out of certain neighborhoods (see Takahashi 1998).

36. Considered a monumental design failure, the high-rise Pruitt-Igoe housing project was destroyed by order of the U.S. Department of Housing and Urban Development and the local housing authority. Many commentators mark this as the end of America's experimentation with "modernist" models of public housing provision.

The Illusion and Necessity of Order

Toward a Just City

On September 11, 2001, University of Arizona Professor of Law Bernard E. Harcourt published an essay on the op-ed page of *The New York Times* that, no doubt, has been all but forgotten in the ensuing maelstrom. But it is an essay worth considering for a moment. September 11 was to be primary election day in New York. At stake were the party nominations for mayor—for the candidates who would run to succeed Rudy Giuliani. Harcourt's essay was titled "The Broken-Window Myth" (Harcourt 2001b), and it drew on research he recently published in his book *Illusion of Order: The False Promise of Broken Windows* (Harcourt 2001a). This research, and other works like it (Sampson and Raudenbush 2001), shows that "broken-windows" policing likely does not reduce crime. Other factors—the waning of the crack epidemic, an improved economy, and simply more police—seem to do a better job of explaining the drop in crime that the 1990s witnessed in most cities in the United States. Harcourt argued that "the best social science research . . . suggests that rather than leading to serious crime, disorder—like crime—is caused by conditions like poverty and a lack of trust between neighbors" (Harcourt 2001b, A23). Not only does broken windows policing, and the whole ideological apparatus that has grown up around it, do nothing to alleviate poverty (indeed, I would argue that one of its main effects is precisely to turn attention away from the need to address poverty), but it can positively increase the level of distrust. As Harcourt (2001b, A23) concluded, broken windows policing "diminishes trust between the police and the community, violates basic rights and scapegoats the homeless and other people we deem disorderly." The missed opportunity of the mayoral primary campaigns, Harcourt suggested in this essay, was that a serious discussion on broken windows policing was never broached during the mayoral campaign. All the candidates had endorsed the Giuliani administration's theory of policing "in some form" (even if some candidates worried about violations of civil liberties) (Harcourt 2001b, A23).[1]

Subsequent to the September 11 attack on New York City, such a debate is even more unlikely. Mayor Giuliani has been lionized for his handling of the crisis, and any criticism of his previous 8 years of maladministration in the city is exceptionally muted. But such a debate, as I hope the previous two chapters has indicated, is desperately needed, for it raises still valid questions—no matter how much the attacks of September 11 may have shifted our sense of what cities do, what they are for, and how the right to them may have been changed. Those questions are: Just what *sort* of order is best for the city? What sort of order will promote the most just city—and for whom? Who will have the right to the city? Who, in the language of both Lefebvre and Bunge, will be allowed to *inhabit* the city? These questions do not at all ask whether there should be any "order" whatsoever, but they ask the more political and more important question of whom that order should serve and by whom it should be governed.

These are the same questions, of course, that animated Matthew Arnold. The riot in Hyde Park in 1866 seemed to portend a frightening, lawless future, and stringent controls on behavior, and on the use of public space, seemed necessary if "anarchy" was to be kept at bay and the possibility of "culture" doing its work was to be maintained. The texture of fear in public spaces has changed during the intervening years, and both the design and legal approaches to order have advanced. But now we are asked to contemplate, along with all the "security experts" who have come to the fore in the wake of 9/11 (see the Introduction), whether it is realistic anymore to maintain even an *expectation* of public space—an expectation of a space in which one can both be public and anonymous, in which the play of politics can be given relatively free reign. Isn't public space a "luxury" we simply can no longer afford?

And yet, at the same time, public spaces, and especially spontaneous gatherings in them, have been an enormous solace in the wake of the September 11th attacks. People have sought out each other in public space. Vigils have been held, and protests—yes, even protests—organized. But if such gatherings reinforce the necessity of urban public space, they obviously do little to address the problems that broken windows policing and quality-of-life initiatives arose to either meet or divert attention from. These problems remain urgent. People still *live* in public spaces, and they are still dying there too. The issues of urban alienation that the Berkeley Free Speech and People's Park movements confronted remain just as important as they were before September 11. What is the connection between these two issues, the issue of homelessness and "quality of life," and the issue of urban alienation for which struggled-over public space can be a (partial) solution? Perhaps that relation can best be seen, in the negative, in the dream of the completely controlled city that the *New York Times*'s security experts described in the wake of the attacks (Barstow 2001). Here is a city in which everyone is known and every movement is predicted. It is a city in which nothing goes unnoticed and in which all behavior is properly scripted. In such a city, as in a city in which all property is private, there simply is no room for homeless people. Indeed, the homeless become even more threatening, for one of the problems that homeless people pose is that they simply are not predictable. They may hide all manner of pathologies. A sizable street population, even more, may provide a cover for unsavory characters—just like that "criminal element" in People's Park that so worried the University of California and the Telegraph Avenue merchants and which then became the pretext for attempts to crack down on the whole of the homeless population. Nor can there be any room for the messy spontaneity of politics, for the unorchestrated play of difference that Iris Marion Young and others assert is so necessary to the development of a just public sphere. When all is controlled, there can simply be *no* right to the city, unless, of course it is for you—for your desires, for your interests, maybe even for your needs—that the city is controlled in the first place.

In this sense the vision of the security experts is in fact the same vision as those who promote broken windows policing and other "quality of life" initiatives: the broken windows thesis says that all homeless people must be criminalized—made suspect—because of the propensity of *other people* to commit crimes. Similarly, the vision of the city promoted by security experts indicates that *all* people in public will need to be made suspect since there might be some in the city who are terrorists (or even lesser criminals). The dream of the perfectly ordered city, then, is exactly the dream in which the city is fully alienated from its residents, placed under total control: it is an authoritarian, even totalitarian, fantasy. Broken windows policing—criminalizing the many so as to hopefully deter the crimes of a different few—does not seem to work on its own terms. What chance is there that the perfectly secure city, a city that operates on exactly the same principle as broken windows policing, will fare any better? The world promoted by the security experts will likely only create the *illusion* of order while at the same time implementing an urbanism that is as alienating as it is controlling. Will the wholesale criminalization and rounding up of people like that which Santa Ana undertook in the early 1990s—but now in the name of security as well as "comfort"—seem a small price to pay?

What other sorts of "order," we need to ask, are imaginable in the "post-9/11" world? What other sorts of order are necessary? And who gets to make that decision?

SPACES OF JUSTICE

If, as Mike Davis (1992) avers, "[t]he universal consequence of the crusade to secure the city is the destruction of any truly democratic space," then the search for a *democratic* order of public space must begin by questioning that crusade to secure the city. The first question to be addressed is in whose interest the city is being secured. In the wake of the terrorist attacks of September 11, the answer to that question must seem obvious: it is being secured in *our* interest. And yet, as the historical and geographical processes examined in this book make clear, the crusade to "secure the city" is not new, and every attempt to reorder the city has served *particular* interests.

Those particular interests, however, have always been contested. Certainly workers have contested laws restricting their right to speak or to picket in public space. And as we have seen in detail, student activists in Berkeley likewise found it necessary to assert their right to speak as a means of asserting their right to politics—their right to transform the alienating political structures of the day. The only way to transform, and even to overthrow, the *order*—and hence the interests encapsulated in that order—has been to defy that order, to break laws, to act without proper decorum. Not only have workers found this, of course, but so too did black civil rights activists and women and gay liberationists in the 1950s, 1960s, and 1970s. In each case, taking to the streets and overthrowing the normative order the streets represent—an order marked by racism, by sexism, and by homophobia—have been crucial to advancing the cause of justice. This process—this taking of space— has often—indeed, I will say has *always*—been contentious; it has ever been a struggle. As Iris Marion Young (1990, 240), has argued, "political theorists who extol the value of community often construe the public as a realm of unity and mutual understanding, but this does not cohere with our actual experience of public space." In public space, "one always risks encounter with those who are different." And those who are different might necessarily be struggling for a place in that public.

But concerted movements in and over public space seem less pertinent in the case of homeless people. Yet, in fact, homeless movements have historically been, and remain, quite important. At the turn of the 20th century the various "armies" that marched on the national and state capitals—like Coxey's Army in 1893 (Schwantes 1985) and Kelly's Army in San Francisco in 1914 (McWilliams 1971 [1939], 164–166; Parker 1919)—were made up of considerable numbers of what we would now call homeless people (people living in run-down flop houses, living rough in tramp jungles, etc.).[2] And they were met by every bit as much violence as were the IWW around World War I or the People's Park activists of the 1960s. More recently, encampments of homeless people—Justiceville in Los Angeles, Camp Agnos in San Francisco (named after an early 1990s mayor), the camp-out on the Santa Monica City Hall steps discussed in Chapter 5, and the encampment eventually cleaned out of Tompkins Square Park, like the Center for Creative Non-Violence's "Reaganville" moved out of Lafayette Park (see Chapter 6)—have been deeply political statements. They have been important and striking commentaries on the current urban "order." And they have been a loud "cry and demand" for a new order.

One of the most interesting current interventions into the "order" of public space—the campaign to secure the city—is Tempe, Arizona's "Project S.I.T." (*http://www.public.asu.edu/~aldous/*). Project S.I.T. was created "to study, analyze, and challenge sidewalk ordinances and other 'public behavior' laws that aim selectively at homeless/street people and their right to exist in public places." Project members engage in civil disobedience (staging sit-ins on Tempe and other Arizona sidewalks, thereby updating that old tradition of disobedience that runs through the labor activists of the first part of the 20th century and the civil rights and student activists of the 1950s and 1960s), legal action, education, and outreach. Project S.I.T. won an important victory in early 2000 when it secured a court decision overturning Tempe's "anti-sitting" law. An appellate court later reversed that decision, but the case is still unsettled, as a new round of appeals has been filed. Project S.I.T.'s aim, like that of lawyers for homeless people in New York and Santa Ana, is not really (or not only) to secure the right to sit on a sidewalk or to sleep in a park, which as we have seen is a pretty narrow right indeed. Rather, it is to contest two related issues: the privatization of public space and the lack of decent and affordable housing. Project S.I.T. members have found that many of the apparently public sidewalks of downtown Tempe have been deeded to developers and business owners. With private, rather than public, property rules in effect, therefore, people can be removed for entirely arbitrary reasons, that is, without cause. Project S.I.T. is devoted to exposing this wholesale transfer of the public realm into private hands and to reversing it—to asserting a right to the city for all and not just for the cappucinoed few. Simultaneously the Project continually highlights the fact that there is not a single homeless shelter in the city of Tempe. In such a circumstance, the enforcement of no-sitting and no-sleeping laws is simply perverse; it quite fundamentally denies the right of some people to *inhabit* the city. Project S.I.T. is working to implement a more just vision of order on the streets of Tempe.

There is a specific way to think about these twin issues of privatization and alienation, on the one hand, and the right to inhabit and the right to the city, on the other, as they relate to public space. As Richard Van Deusen (2002) convincingly argues, public space in the city is a barometer of "justice regimes." By this he means that the morphology of exclusion and access, of power and marginalization, and of struggle and the possibility of representation encapsulates the existing structure of social justice. On the one hand, then, the very fact that so many people are forced, by the current structure of the economy and by public policy, to sleep in public—to "colonize" public space—speaks volumes about the current real social relations within which we live. On the other hand, the fact that we then criminalize those who are so forced to live, that we seek to order public space such that there is simply no room for those evicted from the housing market and forced to live on the streets,[3] is positive testimony about where true power resides—and of what justice consists. Similarly, the fact that we have so privatized public space—encapsulating it within malls, subjecting it to special Business Improvement District regulations, deeding it to private corporations— that it is all but impossible to effectively picket (no matter what current jurisprudence might say about the right to picket) speaks volumes about what we as a society think of the rights of labor. The fact that there are some in society who will nonetheless fight to expand peoples' rights to public space—"gladiators for liberty," as George Will and other "little Arnolds" would sneeringly call them—also speaks volumes. Whatever our *ideals* of justice may be, these facts, these struggles, and their reification in public space are more important.

And yet our ideals of public space *are* important, as Chapter 4 suggested. They frame *how* we struggle and to what end. They guide our desires for and uses of public space. To the degree that we seek to make public spaces spaces of justice, then we work to that degree to implement a particular ideal—or what I call a "vision"—of public space, an ideal based on a commitment to reconstructing the right to the city such that it becomes a reality not for capital, and certainly not just for the wealthy or the suburban shopper, but for all. Yet, as Young (1990, 241) correctly argues, "an ideal can only inspire action for social change if it arises from possibilities suggested by actual experience." While much of this book has been geared toward examining what Habermas (1989) might have called the "structural transformation of public space"—that is, the legal and other practices, often heavily dominated by capital and the forces of a restrictive, intemperate order, that have shaped and reshaped public space in America—its subtext has been precisely that it is *struggle*, and not just the domineering actions of the powerful, that truly shapes public space.

There have been victories. The IWW was in fact quite successful in its many free speech fights; black civil rights activists have done far more than make downtown an acceptable place for an African American person to be homeless, as Ellickson sickeningly suggests (see Chapter 6), and the student activists in Berkeley quite radically transformed what the public space of an American campus could be.[4] The very survival of People's Park in the face of so many pressures to "reform" it is testimony to the ability of ongoing struggle to maintain a certain vision of public space.

This "vision"—and its opposite—can be understood in slightly different spatial terms. Public space is, in some senses, a utopia. The ideal of an unmediated space can never be met[5]—nor can the ideal of a fully controlled space in which the public basks in the splendor of spectacle but is never at any sort of "risk." In *Spaces of Hope*, David Harvey (2000) argues that utopian visions can be divided into two classes. First are "utopias of spatial form." These are the traditional utopias— Thomas More's original Utopia and its many descendants—that seek to specify a spatial form, an arrangement of people and things on the earth, that is fully just, even "happy." Once the proper spatial form is specified, history can come to an end, as a proper and just arrangement of relationships and things has been achieved. The dream of a perfectly secure city, the dream of a risk-free public space in which the consuming classes do not have to be troubled by the sight of disheveled homeless people, like the dream of an orderly culture that has been inherited from Matthew Arnold, is just this sort of utopia. And, as with Arnold (or for that matter Ellickson or the architects of the broken window thesis), exceptionally restrictive codes, covenants, and laws have to be implemented to guarantee the success of the spatial order presumed to be made manifest in the space itself.[6]

The second form of utopia is what Harvey calls "utopias of social process." These range across the political spectrum but can be seen perhaps most clearly in Marx's work, according to Harvey (2000, 174– 175). On the one hand, Marx exposed the utopianism of free market ideologies. The "perfect" free market is a utopia of social process: "give free markets room to flourish, then all will be well with the world" (Harvey 2000, 175). The assumption is that, once the process is set into motion, utopia will result. On the other hand, Marx himself proposed an alternative utopia of social process, a utopia in which class struggle continually upsets the social order until that moment when the oppressive power of expropriators is itself expropriated. Utopia of process is a much more complex form of utopia than utopia of spatial form. But it only raises what Harvey (2000, 177) sees as the key issue, which is the issue of just "what happen[s] when the utopianism of the process comes geographically to earth," a question made obvious even in the earliest years of the Russian revolution (to say nothing of the American: it was, after all, within a decade of the ratification of the Constitution that Congress passed the Alien and Sedition Acts). "The upshot," Harvey (2000, 179) argues, "is that the purity of any utopianism of process inevitably gets upset by its manner of spatialization."

We have, then, a dialectic. Utopias of spatial form (of which the dream of a perfectly ordered public space as encapsulated in anti-homeless or anti-picketing laws is one) are "upset" by the social processes that must be put in place to make utopia reality. And any utopia of social process (of which the dream of a fully democratic and inclusive public space is surely one) must inevitably be "upset" by the spatial form that it takes. Social struggle—the sorts of struggles that underlie the analyses of the preceding chapters even when all the weight of those analyses seems to come down on the side of struggle's opposite, the implementation of a repressive social order—remains critical to the actual structuring and shaping of social justice. And this is why, following Van Deusen (2002), public space *must* be understood as a gauge of the *regimes* of justice extant at any particular moment. Public space *is*, in this sense, the space of justice. It is not only the space where the right to the city is struggled over; it is where it is implemented and represented. It is where utopia is both given spatial form and given lie to. Utopia is impossible, but the ongoing struggle toward it is not.

Despite its impossibility, Harvey (2000, 196) nonetheless argues for the importance of utopia. He asks, "How, then, can a stronger utopianism be constructed that integrates social process with spatial form?"[7] Harvey argues, rightly, that any answer to this question "has to face up to the materialist problems of authority and closure." His discussion is worth repeating:

> Closure (the making of something) of any sort contains its own authority because to materialize any one design, no matter how playfully constructed, is to foreclose, in some cases temporarily but in other instances relatively permanently, on the possibility of materializing others. We cannot evade such choices. The dialectic is "either/or" not "both/and." What the materialization of utopianism of spatial form so clearly confronts is the problem of closure and it is this which the utopianism of the social process
> ¹¹ ⁣dᵤₙ ₘᵢₙₙₙₙₗᵤ ₘᵢₙ dₙₙ. (Harvey 2000, 106)[8]

The materialization of *order*, in other words, is as inevitable as it is necessary. While the violent struggles of the IWW, the seemingly radical demands of the FSM, the taking of People's Park, and the myriad activities of all those "gladiators of liberty" who seek to create a more just space for homeless people in the city might seem the antithesis of "order," in fact the struggle is exactly over what sort of order is to be materialized—and what room there will be in this order for a more democratic "process."

At the same time, order must be *contingent*—contingent on social and economic equality, or at least the ongoing push in that direction. This is particularly important because, as Harvey notes, any imposition of order necessarily forecloses other, alternative materializations. And this, then, is exactly why Raymond Williams urges us to return again and again to Hyde Park, for without doing so new processes and new spatialities cannot be emplaced. *Claiming* the right to the city requires never taking that right for granted, never being satisfied with how it is for now "closed," how it is for now "secured," how, for now, utopia has been materialized. *Expanding* the right to the city requires a clear focus on the utopic possibilities, and the dangers, of always seeking to re-open, and thus to reform, public space in the image of a more just urban order.

NOTES

1. Harcourt goes on to explicitly state the obvious point—that the erosion of civil liberties is an *integral part of* broken windows policing (such policing is based on the erosion of liberties), not an accidental or incidental—and hence correctable—by-product of it. New York was not the only city where broken windows policing was central to a mayoral campaign. One of the candidates for mayor in Seattle during the same election cycle was City Attorney Mark Sidran, who as we have seen is a keen promoter of "quality of life" campaigns. Sidran lost an exceptionally close race, one decided by late-counted mail-in ballots.
2. An excellent recent accounting of the context for these movements (if not for the movements themselves) can be found in Cresswell (2001).
3. See, especially, Deutsche (1996).
4. It is just a shame that so many of those who benefited from the Free Speech Movement are now, as their own children reach college age, some of the most vociferous proponents of a return to *in loco parentis*. On my own campus this is apparent in everything from demands that the faculty monitor the drinking habits of students, to frequent requests to the administration that it report to parents on the political behavior of their children, to calls from parents to faculty asking for mid-term grades for their children.
5. This is the implication of Lefebvre's (1991) argument about the production of space—that all space is always social and so mediated—even if at times that implication gets lost in his overly abstract rendering of the history of space.
6. Harvey (2000, 169–173) uses the rise of "neo-traditional" or "new" urbanism as his primary example of a contemporary utopia of spatial form. New urbanism has drawn on, and fed into, that slightly older utopia of spatial

form, Disneyland, to create what turns out to be a hugely restrictive set of social relationships: this utopia of spatial form is only possible by restricting behavior. The spatial determinism hoped for in design is turned on its head. See, along these lines, Al-Hindi and Staddon (1997). For wider analyses of the stringent regulatory regimes that have been developed to protect the "bourgeois utopias" of suburban America, see MacKenzie (1994) and Fishman (1987).

7. Iris Marion Young (1990, 241) puts the issue in the negative: "the ideal of city life as eroticized public vitality where differences are affirmed in openness might seem laughably utopian. For on city streets today the depth of social injustice is apparent: homeless people lying in doorways, rape in parks, and cold-blooded racist murder are the realities of city life." The question for her is how we construct a city in which domination and oppression are minimized and distributions of goods are more just: this is precisely the question of process and form—the "stronger utopianism"—that Harvey raises.

8. All of this makes the utopia that Harvey proposes as an Appendix to *Spaces of Hope* even more curious: Harvey's vision, in the end, is almost anti-urban in its sequestering of *difference* within communal associations of like-minded individuals.

Postscript (2014)
Now What Has Changed?

One of the many great things about Occupy Wall Street was that it proved me wrong.

A key argument running through *The Right to the City* is that the privatization of public space—its enclosure in malls, its overregulation in the interests of nearby merchants and homeowners at the expense of marginalized populations like the homeless, and the proliferation of privately owned publicly accessible spaces—eliminates the literal (that is, physical) space for politics, and especially for oppositional or radical political struggle. If Occupy Wall Street did not entirely destroy this line of reasoning, it did rip some pretty big holes in it. After all, Zuccotti Park was a quintessential POPS (New York City's term for Privately-Owned Public Spaces): a privately managed public park that had been built and rebuilt in exchange for zoning variances. While city rules governed some aspects of access—it had to be open around the clock, for example—Brookfield Properties, the owner of the space, had wide leeway in establishing and enforcing rules of use. People do not have a fundamental right to use parks like Zuccotti politically, as they putatively do in public streets and parks. However hemmed in and limited the right to speech and protest in publicly owned public space may be— and there is plenty of evidence in *The Right to the City* that political rights in public-public spaces are indeed quite hemmed in and limited—there still remains in the United States a fundamental, if only ever partial, constitutional right to use streets and parks "for the purpose of assembly, communicating thought between citizens, and discussing public questions" as the Supreme Court put it in *Hague v. CIO* (1939) (discussed in Chapter 2). Such rights do not exist in shopping malls (cf. Staeheli and Mitchell 2008), and they are often barely a hope in POPS, both in terms of how they are legally structured, and particularly in how they are designed and governed (Miller 2007).

Any social struggle requires *space*. Occupy Wall Street took and re-made just the kind of space—privately-owned and controlled—I thought would be least likely. This was fantastic. OWS took a park and shook to the core its meaning as well as the fundamental strictures governing it. In doing so it created not just a new kind of public space—and it definitely created that—but a space for a new kind of world. The Occupy movement upended relations and regimes of property, even as they did not disappear altogether. For example, Occupy Syracuse (in upstate New York) set up camp on a privately owned open plaza adjacent to (and part of the same property as) an office tower housing a branch of the Chase bank. Had the Occupiers camped on the more comfortable, grassy, city-owned property next to this plaza they could have been quickly evicted by city police using either protest permit regulations (Occupy Syracuse did not have or want a protest permit) or anti-camping laws. But since Occupy Syracuse occupied private property, city police had to obtain a complaint from the property owner or tenant before they could be evicted for trespassing. At least in the early weeks of the Occupation, the local managers of Chase were reluctant to file such a complaint fearing a backlash, given the popularity of the movement. In Syracuse, as to a fair degree in New York City, existing property laws aided the Occupy movement, even as the property regime was being upended and reworked and new definitions of what constituted public space were being invented. All of a sudden, capitalist private property was protecting the wrong thing, the wrong people, the wrong interests.

So I wasn't the only one who was shown to be wrong by Occupy Wall Street. "Rights skeptics," like those discussed in Chapter 1, were also shown to be wrong, at least partially. Such skeptics tend to worry that "rights talk," and, especially, the actual practice of rights typically protect the wrong interests: they too often protect the rights of the more over the less powerful. But all of a sudden, and for a time, rights were reworked to protect the less powerful—the 99% and especially their representatives camped out in Zuccotti Park and hundreds of other privately owned public spaces around the country. Rights, as I argue in this book, are an institutionalization of power. Marx was correct when he argued that between equal rights, force decides; but that force, we saw with OWS, can be a malleable thing, claimed and put into practice, buttressed and made stronger, by organized masses as much as by police, other agents of the state, or the capitalist class.[1] Yet such organization of force—in this case represented in the sheer mass of bodies as well as the joy attendant upon struggling for and inventing a new kind of world— inevitably calls up a reaction.

When it came, the reaction, from Oakland and Los Angeles to New York and even Syracuse, was swift and brutal. It tended to follow a particular template. First, media stories of filth and danger, of illegitimate disruption to everyday life, of the effrontery of those intent on monopolizing public space and thereby excluding the "legitimate public" began to replace the initially sympathetic coverage of the movement, its tactics, and its demands. City governments, with Mayor Michael Bloomberg's New York leading the charge, began to openly worry about the dirt of the encampments, laying the groundwork for the necessity of cleaning them (out). They often voiced these concerns in terms of dangers to the Occupiers themselves, sometimes trotting out sensational examples of crime and violence to buttress their case. Worries about the economic impact of the encampments intensified, echoing through the media. The steady drumbeat of "something must be done" quickened. Eventually a city would announce, always reluctantly, that it was going to have to clear out the Occupation. A deadline by which all Occupiers had to decamp was broadcast. Police and bulldozers, the city announced, were readied to move in; armies of sanitation workers with dump trucks and power washers were mobilized. Occupiers and their supporters rallied to defend the encampment. First they reorganized their own sanitation and internal policing efforts to dispel worries about cleanliness and safety. When the deadlines were not rescinded, whatever their efforts, they massed at the park to fend off the cops and dump trucks. The cops and sanitation workers inevitably failed to show up. The city seemed to back down. The supporters and Occupiers celebrated. The supporters went home. Two or maybe three nights later, after midnight, when few were up and about, the cops finally moved on the Occupation in a great show of force, riot shields and all, rousting people out of their tents and sleeping bags, piling belongings in great heaps, blasting the pavements with their power washers, positioning front loaders to dump the mass of stuff—books and tents, sleeping bags and camp stoves, folding tables and protest signs—into dumpsters rushed to the scene. Whatever resistance the Occupiers could quickly call into existence was easily routed. Innumerable commentators in the media regretted that such an action was necessary, but after all public spaces are meant for the use of all the people, not just political activists whose message was never properly focused, and there was, anyway, a real and present danger to the Occupiers themselves, to nearby residents and merchants, to the economic well-being of the city.

We've heard this story before. The template was in fact developed over the course of America's now 30-year-long war against homeless people who occupy public space. It is exactly the strategy used against homeless tent city after homeless tent city, from St. Petersburg, Florida, and Camden, New Jersey, to Ontario, Fresno, and Sacramento, California (Mitchell 2013). Initial sympathy for the plight of the homeless gives way to demands that something must be done; deadlines are announced and then ignored; raids eventually come days later, in the middle of the night, and replete with wanton destruction of homeless peoples' things. Tent cities become targets for brutal destruction, especially when they become politically organized, when they show that homeless people can develop ways of living outside the shelter and charity system, that they can collectively resist what I call in this book "the annihilation of space by law." Like Occupy, tent cities sometimes become a potent political force, a reorganization of the structures through which "force decides," and it is for that reason that—for the capitalist state and for bourgeois society alike—they must be destroyed.

One of the infirmities of the Occupy movement, whether in Oakland or New York—or beyond to London and Frankfurt—is that it never fully came to terms with the presence, needs, desires, and politics of homeless people. The media sometimes used this against Occupiers: their kitchens too often gave food away to homeless people who had no interest in the movement, the papers often announced; homeless mentally ill people made the encampments unsafe for all and Occupiers were irresponsible when they did not push them out. But it was a difficult issue. Homeless people were in many ways archetypal 99%ers, evicted from the economy and most of the spaces of the city, forced to live out their lives either in public space or in city and charity shelters within which they had little or no autonomy and few rights. And yet it was sometimes hard to be sympathetic to this or that individual home-

less person, or this or that group of the homeless who joined the encampments. The US's disinvested and privatized mental health system,
together with its failed experiments with deinstitutionalization, means
that rates of mental illness are highly inflated among homeless populations. This *is* dangerous, to homeless people and others alike, and it
means that the presence of significant numbers of homeless people in
Occupy encampments posed serious challenges for self-governance.
More than that, though, Occupy tended not to learn the lessons of
homeless people's tent cities, to not see how the latter were the template
for the former. Occupy was in essence recreating Tent City, though with
a more forceful politics and much wider public involvement. It's not
surprising that the same police tactics were used against Occupy as have
long been used against Tent City. In both cases cities seem to have a fundamental need to retake public space—space that has been made public
in a new way, like New York's Zuccotti Park or St. Petersburg's New
Hope City homeless encampment.

There have been plenty of examples that Occupiers could have
paid attention to, and thus had a better sense of what was coming. In
the decade since *Right to the City* was published, the *legal* war against
homeless people, which has been fought through the writing and selec-
tive enforcement of anti-homeless laws that seek to make it impossible
for homeless people to *be* in public space (Chapters 5 and 6), has not
only continued unabated, but intensified (as the annual reports of the
National Law Center on Homelessness and Poverty make plain). And it
has been joined by a full frontal assault on the places homeless people
carve out to nonetheless find a way to survive—such as the assaults on
tent cities—as well as on those from outside the sanctioned charitable
sector who seek to work with homeless people to make their lives better
(Mitchell and Heynen 2009). Now even the giving away of food to hungry people has been criminalized in many jurisdictions.[2] Such an ongoing assault on homeless people in public space, even as the US economy
remains so ruthlessly adept at producing homelessness, gives the lie to
the rather ridiculous claim arising in some scholarly quarters—often
against the arguments put forward in this book—that because some
people care about the lives of homeless people, because some charitable
organizations provide beds and care and a large degree of compassion,
we *do not* live in what I call the "post-justice" city or what Neil Smith
(1996) so famously named the "revanchist" city (e.g. Cloke, May, and
Johnsen 2010). Anyone who knows the history of the managing and po-

licing of homeless people in the city knows that charity often goes hand-in-hand with the most revanchist of policies. It's the velvet glove covering the mailed fist. As the Industrial Workers of the World exposed a century ago charity was (and is) made possible by the same class formation that demanded a ban on the Wobbly's rights of free speech in San Diego, Spokane, and Seattle. The point is to contain and control the potentially disruptive and disorderly. Anti-homeless laws, as well as the erection of an elaborate shelter system, are meant, however contradictorily, to do just that, no matter how much they might also be meant to provide a homeless person with a more or less safe bed for a night.

As Chapters 3 and 4 show, Berkeley, California, and People's Park in particular have been prime battlegrounds in the war over the presence of homeless people in public space. In the years since the conflicts detailed in *The Right to the City* took place, struggle in and around People's Park has waxed and waned, but never disappeared. The volleyball courts that led to the 1991 rioting have long since been torn up and replaced with waterlogged sod, but the basketball courts and bathrooms remain, as does the Free Stage. On December 28, 2011, not long after the Occupy encampments had been cleared out of most US cities (including the quite violent destruction of nearby Oakland's militant encampment in Frank Ogawa Plaza in front of city hall), the University of California once again sent bulldozers into People's Park. The move was focused on the west end of the park behind the Free Stage, an area of community gardens, tall trees, winding paths, and, according to the University of California, rats. Some of the gardens ripped up dated to 1979 when activists tore out a small asphalt parking lot UC had built on the site. The bulldozers this time also ripped up a pergola that had been built out of the net-posts of the destroyed volleyball courts. Besides eliminating rat habitat, the university claimed its actions were necessary to improve views for students moving into a soon-to-be-completed dorm across the street. Despite a promise to do so, the university did not consult with the People's Park Community Advisory Council before sending in the bulldozers (which seems to have been a long-delayed first-step in implementing a new design strategy for the park that many activists had objected to even as the Advisory Council had been a party to its development).

This time, though, the arrival of the bulldozers and fence-builders was met with almost no opposition. The lack of mobilization suggests, perhaps, that after more than four decades of struggle and stasis, People's Park is ripe for change. The Telegraph Business Improvement District certainly thinks so, arguing, once again, that the Park was mostly "a haven for drug dealing, drug and alcohol abuse and the various anti-social and criminal behaviors they induce" (Patterson 2012). What is missing from this list, compared to earlier pronouncements by Telegraph merchants (see Chapter 4), is the presence of the homeless. Perhaps they have learned not to demonize an already marginalized class. More likely they are pretty good at assessing local opinion. Only two months earlier Berkeley residents went to the polls and soundly defeated Measure S, a proposed "sit-lie" law—that is, a classic anti-homeless law that would have banned sitting or lying on sidewalks in commercial areas in commercial districts (with exceptions for medical emergency and for someone sitting in a wheel chair, on a permanent public bench, at a street event for which a permit had been received, or in outdoor café seating). It was the first time since 1994 that voters in any city in the US rejected a ballot measure criminalizing homelessness.

It would be nice to think that this vote indicates, finally, that the full frontal assault on homeless people in public space is beginning to fade. Coupled with Occupy Wall Street's stunningly successful *taking* of privately owned public space (and Occupy Oakland's similarly successful taking of public-public space), it might indicate, as I suggest in *The Right to the City,* that pronouncements of *the end* of public space are premature, and way too simple. What the vote, like Occupy, really represents though is a tremendous organizational effort to take back, to reclaim, to keep open and vibrant public life and the spaces that make it possible. The campaign against Measure S, which was clever in its tactical use of theater in the streets and sidewalks to press home its case, came hot on the heels of sustained protest by UC Berkeley students against mass budget cuts, large tuition rises, significant privatization of public university assets and functions, and, generally, the total selling out of the idea of a public education. Buildings were occupied, the administration's response was chaotic, and Berkeley students joined a wave of student protests and university (and high school) occupations that shocked complacent neoliberal administrators and governments from Glasgow to Santiago. Indeed, students are in the midst of creating a global movement—not yet fully coalesced—of potentially greater transformative importance than the one more or less kicked off by the Berkeley Free Speech Movement fifty years ago (Chapter 3). This student movement intersects with—and sometimes has helped to spark—protests and occupations by *indignados* in Spain; pissed-off Icelanders in Reykjavik; Greeks fighting looting by the Troika; the masses who rose up under the banner of Arab Spring (and all those, like the occupiers of Istanbul's Gezi Park, who have continued to take public space to contest autocratic rule and the usurpation of the city); Right to the City activists in New York, Mexico City, and Istanbul; and soon enough yet more masses of students and citizens across Brazil, each with a million complaints about the privatization of everything and the ransacking of collective, public wealth, collective, public goods, and collective, public spaces by unaccountable banks, corporations, and their paid-for politicians. People across the globe continue to discover again and again that making a new kind of world requires taking space and making it anew as a *public* space.

It requires "going again to Hyde Park," as Raymond Williams argued. Virtual protest is no protest at all (no matter how valuable social media may be for organizing protest). Physical presence, which requires *taking* space, is everything. Space is *made* public—and the meaning of "public" is redefined—through its taking. One of the truly salutary effects of the Occupy movement and all that has surrounded it—from the mass, global protests against the war against Iraq in 2003 to the most recent Brazilian uprisings—is that they have shown that "going again to Hyde Park" is not only necessary but possible. Doing so always calls up ferocious reaction—both the ideological reaction from the cadres of "little Arnolds" always ready to justify the justness of the already powerful, and the police violence these little Arnolds pave the way for and explain away. This is why it is always necessary to go *again* to Hyde Park. This is why it was so deliciously wonderful to be proved wrong by Occupy Wall Street as to where that "Hyde Park" might be: privatized public space *can* be made collective, public, and political, and it must be.

In the decade since *The Right to the City* was first published, the struggle for, in, and over public space—and therefore for the right to the city—has only intensified. In the process new worlds, new spatial forms of justice, can be glimpsed, if too fleetingly, as uprising and occupation after uprising and occupation is beaten back only to arise again in a new form somewhere else. Homeless people remain a bellwether of justice in the contemporary city, and the ways they are policed, as well as the meaning of them being policed, remain closely tied to the fate of urban social movements. I may have been wrong about the possibility for politics in privately owned public space, but I do not think the analysis in *The Right to the City* is at all wrong in asserting that the struggle for public space is vital to understand if we are to understand what might possibly constitute a just city, a just society, a just urban *order.* I hope you will agree that the analyses of the struggles for, in, and over public space in this book retain just as much political and theoretical salience as they did in 2003. I hope, especially, that you will find other opportunities to make the pessimism that accompanies many of these analyses—pessimism about the possibility of radical politics in urban public space—as wrong as Occupy Wall Street did.

NOTES

1. I do not mind romanticizing the Occupy movement a little bit. There is no doubt that it was limited as a movement and in its politics. In many locales it was an overwhelmingly white movement, and seemed centered on the demands of culturally alienated middle class youth rather than the economically alienated majority it claimed to represent. And it certainly always seemed to want to prove the truth of the old joke—"What's the definition of socialism?" "Endless meetings"—only with its anarchism trying to do this one better. That anarchism itself was both its great strength and an Achilles heel. All this is true. But it is also true that Occupy scared the shit out of the ruling class. It scared the shit out of them (and out of many in the comfortable bourgeoisie) in large part because it was in the process of inventing a new way of living, a new way of being, in which "public" and "common"— both rooted in the taking, occupation, and making of public space—operated around, outside, and in the interstices of dominant society that is alienating both culturally and economically.

2. Sometimes this criminalization is outright, as with attempts in Las Vegas and other cities to make it illegal to give away food for free in public spaces. Sometimes it takes the form of new sanitation and public health laws or their stepped up enforcement. Once again the assault on homeless people and their advocates provided a template for the assault on Occupy: a number of cities sought to enforce food safety and public health laws against Occupiers as a means of pushing them out of their occupied public spaces.

References

Ades, P. 1989. "The Unconstitutionality of 'Antihomeless' Laws: Ordinances Prohibiting Sleeping in Outdoor Public Areas as a Violation of the Right to Travel," *California Law Review* 77, 595–628.

Al-Hindi, K. F., and Staddon, C. 1997. "The Hidden Histories and Geographies of Neotraditional Town Planning: The Case of Seaside, Florida," *Environment and Planning D: Society and Space* 15, 349–372.

American Social History Project. 1989. *Who Built America*, Volume 1. New York: Pantheon.

American Social History Project. 1992. *Who Built America*, Volume 2. New York: Pantheon.

Anderson, B. 1991. *Imagined Communities: Reflections on the Origin and Spread of Nationalism*. London: Verso (revised edition).

Anderson, N. 1923. *The Hobo: The Sociology of the Homeless Man*. Chicago: University of Chicago Press.

Arendt, H. 1972. "On Violence," in *Crisis of the Republic*. New York: Harcourt, Brace, and Javanovich.

Arnold, M. 1993. *Culture and Anarchy and Other Writings*. Cambridge: Cambridge University Press, edited by S. Collini.

Atlanta Journal and Constitution 1991. "Harassing the Harassers?" July 15, D1.

Atlanta Journal and Constitution 1993. "Advocates: City Is Trying to Hide the Homeless," July 12, C4.

Auchard, E. 1991. "How Did It Happen? A Protest Diary," *East Bay Express* August 9, 1, 18–23.

Avery, D. 1988/1989. "Images of Violence in Labor Jurisprudence: The Regulation of Picketing and Boycotts, 1894–1921. *Buffalo Law Review* 37, 1–117.

Bahr, H. 1970. *Disaffiliated Man: Essays and Bibliography on Skid Row, Vagrancy and Outsiders*. Toronto: University of Toronto Press.

Bahr, H. 1973. *Skid Row: An Introduction to Disaffiliation.* New York: Oxford University Press.

Bakan, J., and Blomley, N. 1992. "Spatial Categories, Legal Boundaries, and the Judicial Mapping of the World," *Environment and Planning A* 24, 629–644.

Balter, J. 1994. "City's Panhandling Law Becomes a Big Problem for Small Neighborhoods," *The Seattle Times* June 5, B1.

Baltimore Sun 2001. "Bill Would Ban Night Panhandling," June 27, 1A.

Barstow, D. 2001. "After the Attacks, Security: Envisioning an Expensive Future in a Brave New World of Fortress New York," *New York Times* September 16, 1:16.

Baum, A., and Burnes, D. 1993. *A Nation in Denial: The Truth about Homelessness.* Boulder; Westview Press.

Berman, G. 1994. "A New Deal for Free Speech: Free Speech and the Labor Movement in the 1930s," *Virginia Law Review* 80, 291–322.

Bernstein, N. 1999. "Labeling the Homeless, in Compassion and Contempt," *New York Times* December 5, 1:53.

Bittner, E. 1967. "The Police on Skid Row: A Study in Peace Keeping," *American Sociological Review* 32, 699–715.

Blau, J. 1992. *The Visible Poor: Homelessness in the United States.* New York: Oxford University Press.

Blomley, N. 1989. "Text and Context: Rethinking the Law-Space Nexus," *Progress in Human Geography* 13, 512–534.

Blomley, N. 1994a. *Law, Space, and the Geography of Power.* New York: Guilford Press.

Blomley, N. 1994b. "Mobility, Empowerment, and the Rights Revolution," *Political Geography* 13, 407–422.

Blomley, N. 1998. "Landscapes of Property," *Law and Society Review* 32, 567–612.

Blomley, N. 2000a. " 'Acts,' 'Deeds,' and the Violences of Property," *Historical Geography* 28, 86–107.

Blomley, N. 2000b. "Property Rights," in R. Johnston et al., *The Dictionary of Human Geography.* Oxford: Blackwell (4th edition), 651.

Blomley, N. 2004. *Unsettling the City: Space, Property and Urban Land.* New York: Routledge.

Blomley, N., and Clark, G. 1990. "Law, Theory, and Geography," *Urban Geography* 11, 443–446.

Blomley, N., Delaney, D., and Ford, R. (eds.) 2001. *The Legal Geographies Reader.* Oxford: Blackwell.

Blum, V., and Nast, H. 1996. "Where's the Difference? The Heterosexualization of Alterity in Henri Lefebvre and Jacques Lacan," *Environment and Planning D: Society and Space* 14, 559–580.

Blumberg, L., Shipley, T., and Barsky, S. 1978. *Liquor and Poverty: Skid Row as a Human Condition.* New Brunswisk, NJ: Rutgers Center of Alcohol Studies.

Boston Globe 2000. "Fewer Homeless Died on Streets of City in 1999," March 31, B1.

Boudreau, J. 1991. "The People Grudgingly Give In on Park," *Contra Costa Times* August 2, A3.

Boyer, C. 1992. "Cities for Sale: Merchandizing History at South Street Seaport," in M. Sorkin (ed.), *Variations on a Theme Park: The New American City and the End of Public Space.* New York: Hill and Wang, 181–204.

Brenner, N. 1997. "Global, Fragmented, Hierarchical: Henri Lefebvre's Geographies of Globalization," *Public Culture* 10, 135–169.

Brisbin, R. 1993. "Antonin Scalia, William Brennan, and the Politics of Expression: A Study of Legal Violence and Repression," *American Political Science Review* 87, 912–927.

Brown, M. 1997. *RePlacing Citizenship: AIDS Activism and Radical Democracy.* New York: Guilford Press.

Brown, M. 2000. *Closet Space.* New York: Routledge.

Bruns, R. 1987. *The Damnedest Radical: The Life and World of Ben Reitman, Chicago's Celebrated Social Reformer, Hobo King, and Whorehouse Physician.* Urbana: University of Illinois Press.

Bumiller, E. 1999. "In Wake of Attack, Giuliani Cracks Down on the Homeless," *New York Times* November 20, A1.

Bunge, W. 1971. *Fitzgerald: Geography of a Revolution.* Cambridge MA: Schenkman.

Bunge, W., and Bordessa, R. 1975. *The Canadian Alternative: Survival, Expeditions and Urban Change.* Department of Geography, York University: Geographical Monographs.

Burress, C. 1999. "Proposal for People's Park Dorms; UC Chancellor Suggests Student Housing for Site," *San Francisco Chronicle* April 8, A1.

Calhoun, C. 1989. "Tiananmen, Television and the Public Sphere: Internationalization of Culture and the Beijing Spring of 1989," *Public Culture* 2: 54–71.

Calhoun, C. 1992. "Introduction: Habermas and the Public Sphere," in C. Calhoun (ed.), *Habermas and the Public Sphere.* Cambridge: Cambridge University Press, 1–48.

Carpignano, P., Anderson, R., Aronowitz, S., and Difazio, W. 1990. "Chatter in the Age of Electronic Reproduction: Talk, Television, and the 'Public Mind.' " *Social Text* 25/26, 33–55.

Chauncy, G. 1994. *Gay New York: Gender, Urban Culture, and the Making of the Gay Male World, 1890–1940.* New York: Basic Books.

Chouinard, V. 1994. "Geography, Law, and Legal Struggles: Which Way Ahead?" *Progress in Human Geography* 18, 415–440.

Cincinnati Enquirer 1995a. "City Council OKs Cracking Down on Panhandling," May 4, B2.

Cincinnati Enquirer 1995b. "DCI to Study Panhandlers, Lead Campaign," April 18, B1.

Cincinnati Enquirer 1995c. "Limits on Begging Before Council Today: Effort to Make New Law Temporarily Fails," May 3, B1.

Cincinnati Enquirer 2002. "Panhandling Law Takes Effect Monday," April 7, B2.

Clark, G. 1990. "The Virtues of Location: Do Property Rights 'Trump' Workers' Rights to Self-Organization?" *Environment and Planning D: Society and Space* 8, 53–72.

Clark, G., and Dear, M. 1984. *State Apparatus*. Boston: Allen and Unwin.

Cleveland Plain Dealer 1994. "Akron Ponders Banning Beggars," July 7, 1B.

Cloke, P., May, J., and Johnsen, S. 2010. *Swept Up Lives? Re-envisioning the Homeless City*. Oxford: Wiley Blackwell.

Cole, D. 1986. "Agon at Agora: Creative Misreadings in the First Amendment Tradition," *Yale Law Journal*, 95, 857–905.

Commercial Appeal (Memphis) 1994. "Other Laws Eclipse City's Panhandling Ordinance," July 24, B1.

Commonwealth of Pennsylvania. 1890. *General Report of the Commissioners Appointed to Revise and Codify the Laws Relating to the Relief, Care, and Maintenance of the Poor in the Commonwealth of Pennsylvania*. Harrisburg: Meyer's Printing House.

Cope, M. 1997. "Responsibility, Regulation, and Retrenchment: The End of Welfare?" in L. Staeheli, J. Kodras, and C. Flint (eds.), Thousand Oaks, CA: Sage Publications, 181–205.

Cosgrove, D. 1984. *Social Formation and Symbolic Landscape*. London: Croom Helm.

Cosgrove, D. 1985. "Prospect, Perspective and the Evolution of the Landscape Idea," *Transactions of the Institute of British Geographers* 10, 45–62.

Cosgrove, D. 1990. "Spectacle and Society: Landscape as Theater in the Pre-Modern and Post-Modern Cities," in P. Groth (ed.), *Vision, Culture, and Landscape*. Berkeley: Department of Landscape Architecture, University of California, 221–239.

Cosgrove, D. 1993. *The Palladian Landscape: Geographical Change and its Cultural Representation in Sixteenth-Century Italy*. University Park: Pennsylvania State University Press.

Cover, R. 1981. "The Left, the Right, and the First Amendment: 1918–1928," *Maryland Law Review* 40, 349–388.

Cox, K., and Mair, A. 1988. "Locality and Community in the Politics of Local Economic Development," *Annals of the Association of American Geographers* 78, 307–325.

Crawford, M. 1992. "The World in a Shopping Mall," in M. Sorkin (ed.), *Variations on a Theme Park: The New American City and the End of Public Space. New York: Hill and Wang, 3–30.

Cresswell, T. 1996. In Place/Out of Place: Geography, Ideology, and Transgression. Minneapolis: University of Minnesota Press.

Cresswell, T. 2001. The Tramp in America. London: Reaktion Books.

Crilley, D. 1993. "Megastructures and Urban Change: Aesthetics, Ideology and Design," in P. Knox (ed.), The Restless Urban Landscape. Englewood Cliffs, NJ: Prentice Hall, 126–164.

Crump, J. 2002. "Deconcentration by Demolition: Public Housing, Poverty and Urban Policy," Environment and Planning D: Society and Space 20, 581–596.

Crump, J. 2003. "The End of Public Housing as We Know It: Public Housing Policy, Labor Regulation, and the U.S. City," International Journal of Urban and Regional Research 27, 179–187.

Czerniak, J. (ed.) 2002. CASE: Downsview Park Toronto. Cambridge: Prestel.

Daniels, S. 1993. Fields of Vision: Landscape Imagery and National Identity in England and the United States. Princeton: Princeton University Press.

Daniels, S., and Cosgrove, D. 1993. "Spectacle and Text: Landscape Metaphors in Cultural Geography," in J. Duncan and D. Ley (eds.), Place/Culture/Representation. London: Routledge, 57–77.

D'Arcus, B. 2001. Marginal Protest and Central Authority: The Scalar Politics of the Wounded Knee Occupation. Unpublished PhD Dissertation, Department of Geography, Syracuse University.

Davis, M. 1990. City of Quartz: Excavating the Future in Los Angeles. London: Verso.

Davis, M. 1991. "Afterword—A Logic Like Hell's: Being Homeless in Los Angeles," UCLA Law Review 39, 325–332.

Davis, M. 1992. "Fortress Los Angeles: The Militarization of Urban Space," in M. Sorkin (ed.), Variations on a Theme Park: The New American City and the End of Public Space. New York: Hill and Wang, 154–180.

Dear, M., and Wolch, J. 1987. Landscapes of Despair. Princeton: Princeton University Press.

Debord, C. 1994 (1967). The Society of the Spectacle. New York: Zone Books (translated by D. Nicholson-Smith).

de Certeau, M. 1984. The Practice of Everyday Life. Berkeley: University of California Press.

Dees, J. 1948. Flophouse. Francestown, NH: M. Jones Co.

Delaney, D. 1998. Race, Place, and the Law. Austin: University of Texas Press.

Delaney, D. 2001. "Making Nature/Making Humans: Law as a Site of (Cultural) Production," Annals of the Association of American Geographers 91, 487–503.

Denver Post 1999a. "Colorado Springs Measure Targets Street Beggars," November 2, B5.

Denver Post 1999b. "Homeless Deaths a National Concern: Rising Housing Costs Mean More People at Risk," November 13, B1.

Denver Post 2000a. "City to Rein in Panhandlers," May 16, B1.

Denver Post 2000b. "Service Recalls Homeless Victims: Names of 58 Dead Read at Remembrance Vigil," December 22, B2.

Derrida, J. 1992. "Forces of Law: 'The Mystical Foundation of Authority,' " in D. Cornell, M. Rosenfeld, and D. Carlson (eds.), *Deconstruction and the Possibility of Justice*. New York: Routledge, 3–67.

Deutsche, R. 1990. "Architecture of the Evicted," *Strategies: A Journal of Theory, Culture, and Politics* 3, 159–183.

Deutsche, R. 1992. "Art and Public Space: Questions of Democracy," *Social Text* 33, 34–53.

Deutsche, R. 1996. *Evictions: Art and Spatial Politics*. Cambridge, MA: MIT Press.

Di Rado, A. 1994. "Appeals Court Voids Santa Ana Ban on Camping by Homeless," *Los Angeles Times* February 4, A27.

Dolan, M. 1995. "State Upholds Tough Homeless Law," *Los Angeles Times* April 25, A1.

Domosh, M. 1998. " 'Those Gorgeous Incongruities': Polite Politics and Public Space on the Streets of Nineteenth Century New York City," *Annals of the Association of American Geographers* 88, 209–226.

Dorgan, M. 1985. "Hippies Moved from Street to Berkeley Dump," *San Jose Mercury-News* January 31, B12.

Draper, H. 1966. *Berkeley: The New Student Revolt*. New York: Grove Press.

Dubofsky, M. 1988. *We Shall Be All: A History of the Industrial Workers of the World*. Urbana: University of Illinois Press (2nd edition).

Duncan, J., and Duncan, N. 2001. "The Aestheticization of the Politics of Landscape Preservation," *Annals of the Association of American Geographers* 91, 387–409.

Editors of the *California Monthly*. 1965. Chronology of Events: Three Months of Crisis. *California Monthly* (February), reprinted in S. Lipset and S. Wolin (eds.), *The Berkeley Student Revolt*. Garden City: Doubleday, 99–198.

Ellickson, R. 1990. "The Homeless Muddle," *The Public Interest*, Spring, 45–52.

Ellickson, R. 1991. *Order Without Law: How Neighbors Settle Disputes*. Cambridge: Harvard University Press.

Ellickson, R. 1996. "Controlling Chronic Misconduct in City Spaces: Of Panhandlers, Skid Rows, and Public Space Zoning," *Yale Law Journal* 105, 1165–1248.

Eng, L. 1990a. "Santa Ana Homeless Suit Settled; Law: The City Council Agrees to Pay $50,000 to 17 People in a Case Involving Confiscation and Discarding of Their Personal Property During Cleanup Sweeps," *Los Angeles Times* February 6, B1.

Eng, L. 1990b. "Santa Ana Police Chief Vows to Continue with Sweeps; Civil

Rights: Despite Questions of Legality, He Defends the Roundup of 64 Homeless Men at the Civic Center on Wednesday," *Los Angeles Times* August 18, B5.

Eng, L. 1991. "Cases Dismissed in Santa Ana's Homeless Sweep," *Los Angeles Times* February 7, A1.

Eng, L., and Drummond, T. 1990. "Another Sweep, More Complaints; Crime: Santa Ana Police Conduct Another Roundup in the Civic Center Area That Rights Activists Say Unfairly Targets the Homeless and Illegal Aliens," *Los Angeles Times* August 22, B1.

Falit-Baiamonte, A. 2000. "Identity, Public Space, and Protest: Scales of Challenge," Unpublished Paper, Department of Geography, University of Washington.

Falk, C. 1990. *Love, Anarchy, and Emma Goldman.* New York: Holt, Rinehart and Winston.

Fasanelli, A. 2000. "Note: In Re Eichorn: The Long Awaited Implementation of the Necessity Defense in a Case of the Criminalization of Homelessness," *American University Law Review* 50, 323–353.

Feuer, L. 1966. "The Decline of Freedom at Berkeley," *Atlantic Monthly* 218 (September), 78–85.

Fink, L. 1987. "Labor, Liberty, and the Law: Trade Unionism and the Problem of American Constitutional Order," *Journal of American History* 74, 904–925.

Fishman, R. 1987. *Bourgeois Utopias: The Rise and Fall of Suburbia.* New York: Basic Books.

Foner, P. 1965. *History of the Labor Movement in the United States, Volume IV: The Industrial Workers of the World, 1905–1917.* New York: International Publishers.

Foner, P. 1981. *Fellow Workers and Friends: Free Speech Fights as Told by Participants.* Westport, CT: Greenwood Press.

Foote, C. 1956. "Vagrancy-Type Law and Its Administration," *University of Pennsylvania Law Review* 104, 603–650.

Forbath, W. 1991. *Law and the Shaping of the American Labor Movement.* Cambridge, MA: Harvard University Press.

Foscarinis, M. 1996. "Downward Spiral: Homelessness and Its Criminalization," *Yale Law and Policy Review* 14, 1–63.

Foscarinis, M., Cunningham-Bowers, K., and Brown, K. 1999. "Out of Site—Out of Mind? The Continuing Trend Toward the Criminalization of Homelessness," *Georgetown Journal on Poverty Law and Policy* 6, 145–164.

Frank, T. 2001. *One Market Under God: Extreme Capitalism, Market Populism, and the End of Economic Democracy.* New York: Doubleday.

Fraser, N. 1990. "Rethinking the Public Sphere: A Contribution to Actually Existing Democracy," *Social Text* 25/26, 56–79.

Fyfe, N. (ed.) 1998. *Images of the Street: Identity and Control in Public Space.* London: Routledge.

Fyfe, N., and Bannister, J. 1995. "City Watching: Closed Circuit Television Surveillance in Public Space," *Area* 29, 37–46.

Fyfe, N., and Bannister, J. 1998. " 'The Eyes Upon the Street': Closed-Circuit Television Surveillance and the City," in N. Fyfe (ed.), *Images of the Street: Identity and Control in Public Space.* London: Routledge, 254–267.

Garreau, J. 1991. *Edge City: Life on the New Frontier.* New York: Doubleday.

Gitlin, T. 1993. *The Sixties: Years of Hope, Days of Rage.* New York: Bantam (revised edition).

Gitlin, T. 1995. *The Twilight of Common Dreams: Why America Is Wracked by Culture Wars.* New York: Metropolitan.

Giuliani, R., and Bratton, W. 1994. *Police Strategy No. 5: Reclaiming the Public Spaces of New York.* New York: Office of the Mayor.

Glazer, N. 1992. " 'Subverting the Context': Public Space and Public Design," *Public Interest* 109, 3–21.

Goheen, P. 1993. "Negotiating Access to Public Space in Mid-Nineteenth Century Toronto," *Journal of Historical Geography* 30, 430–449.

Gold, J., and Revill, G. (eds.) 2000. *Landscapes of Defence.* Harlow, UK: Prentice Hall.

Golden, R. 1998. "Towards a Model of Community Representation for Legal Assistance Lawyering: Examining the Role of Legal Assistance Agencies in Drug-Related Evictions from Public Housing," *Yale Law and Policy Review* 17, 527–561.

Gomez, J. 1990. "64 Homeless Men Seized in Santa Ana Police Sweep; Crime: Officers Write Numbers on Arms of Those Arrested; A Civil Rights Proponent Says the Action Smacks of Nazism," *Los Angeles Times* August 17, B6.

Goss, J. 1992. "Modernity and Postmodernity in the Retail Rural Environment," in K. Anderson and F. Gale (eds.), *Inventing Places.* Melbourne: Longman Scientific, 159–177.

Goss, J. 1993. "The 'Magic of the Mall': An Analysis of Form, Function, and Meaning in the Contemporary Retail Built Environment," *Annals of the Association of American Geographers* 83, 18–47.

Goss, J. 1996. "Disquiet on the Waterfront: Reflections on Nostalgia and Utopia in the Urban Archetypes of Festival Marketplaces," *Urban Geography* 17, 221–247.

Goss, J. 1999. "Once-Upon-a-Time in the Commodity World: An Unofficial Guide to Mall of America," *Annals of the Association of American Geographers* 89, 45–75.

Gostin, L. 1988. "Towards Resolving the Conflict," in L. Gostin (ed.), *Civil Liberties in Conflict.* London: Routledge, 7–20.

Gramsci, A. 1971. *Selections from the Prison Notebooks*. London: Lawrence and Wishart (edited and translated by Q. Hoare and G. Nowell-Smith).

Greenberg, K. 1990. "The Would-Be Science and Art of Making Public Spaces," *Architecture et Comportment/Architecture and Behaviour* 6, 323–338.

Gregory, D. 1994. *Geographical Imaginations*. Oxford: Blackwell.

Groth, P. 1994. *Living Downtown: The History of Residential Hotels in the United States*. Berkeley: University of California Press.

Habermas, J. 1974. *Legitimation Crisis*. Boston: Beacon Press (translated by T. McCarthy).

Habermas, J. 1989. *The Structural Transformation of the Public Sphere*. Cambridge, MA: MIT Press.

Haggerty, J. 1992/1993. "Begging and the Public Forum Doctrine in the First Amendment." *Boston College Law Review* 34, 1121–1162.

Halberstam, J. 1993. "Imagined Violence, Queer Violence: Representations, Rage, and Resistance." *Social Text* 37, 187–201.

Hall, S. 1988. *The Hard Road to Renewal: Thatcherism and the Crisis of the Left*. London: Verso.

Hamill, P. 1993. "How to Save the Homeless—And Ourselves." *New York* 26 (September 20), 34–39.

Harcourt, B. 2001a. *Illusion of Order: The False Promise of Broken Windows Policing*. Cambridge: Harvard University Press.

Harcourt, B. 2001b. "The Broken-Window Myth," *New York Times* September 11, A23.

Harris, B. 1988. "Homeless and Their Neighbors," *Oakland Tribune* February 22, B12.

Hartley, J. 1992. *The Politics of Pictures: The Creation of the Public in the Age of Popular Media*. London: Routledge.

Hartman, C. 1987. "The Housing Part of the Homeless Problem," in J. Kneerim (ed.), *Homelessness: Critical Issues for Policy and Practice*. Boston: The Boston Foundation.

Harvey, D. 1973. *Social Justice and the City*: Baltimore: Johns Hopkins University Press (republished 1988. Oxford: Blackwell).

Harvey, D. 1982. *The Limits to Capital*. Chicago: University of Chicago Press (republished 1999. London: Verso).

Harvey, D. 1989. *The Condition of Postmodernity* (Oxford: Blackwell).

Harvey, D. 1990. "Between Space and Time: Reflections on the Geographical Imagination," *Annals of the Association of American Geographers* 80, 418–434.

Harvey, D. 1992. "Social Justice, Postmodernism and the City," *International Journal of Urban and Regional Research* 16, 558–601.

Harvey, D. 1993. "From Space to Place and Back Again: Reflections on the Condition of Postmodernity," in J. Bird, B. Curtis, T. Putnam, G. Robertson, and L. Tickner (eds.), *Mapping the Futures: Local Cultures, Global Change*. London: Routledge, 3–29.

Harvey, D. 1996. *Justice, Nature, and the Geography of Difference*. Oxford: Blackwell.

Harvey, D. 2000. *Spaces of Hope*. Berkeley: University of California Press.

Heirich, M. 1971. *The Sprial of Conflict: Berkeley, 1964*. New York: Columbia University Press.

Heirich, M., and Kaplan, S. 1965. "Yesterday's Discord." *California Monthly* (February), reprinted in S. Lipset and S. Wolin (eds.), *The Berkeley Student Revolt*. Garden City: Doubleday, 10–34.

Henderson, G. 1999. *California and the Fictions of Capital*. Oxford: Oxford University Press.

Herscher, E. 1995. "Berkeley Still Struggling With Anti-Panhandling Law: Legal Ruling Has Blocked Enforcement," *San Francisco Chronicle*, June 5, A13.

Hershkovitz, L. 1993. "Tiananmen Square and the Politics of Place," *Political Geography* 12, 395–420.

Heyman, R. 2001. "Jumping Scales and Sinking Ships: Public Space in 'Seattle,' " Unpublished Paper, Department of Geography, University of Washington.

Hillis, K. 1994. "The Virtue of Becoming a No-Body," *Ecumene* 1, 177–196.

Hoch, C., and Slayton, R. 1989. *New Homeless and Old: Community and the Skid Row Hotel*. Philadelphia: Temple University Press.

Holtz, D. 2000a. "People's Park Sale Hoped For," *San Francisco Chronicle*, July 8, A15.

Holtz, D. 2000b. "Tensions Rise Between Denizens, Advocates in Battle Over People's Park," *San Francisco Chronicle*, July 17, A18.

Hombs, M., and Snyder, M. 1982. *Homelessness in America: A Forced March to Nowhere*. Washington, DC: The Community for Creative Non-Violence.

Hopper, K., and Hamberg, J. 1984. *The Making of America's Homeless: From Skid Row to the New Poor*. New York: Community Service Society.

Horn, M. 1989. "Berkeley: The Young Radicals Are Now Middle-Aged and Middle Class, Yet the '60s Ethic Endures," *U.S. News and World Report* 18 December, 59–60.

Houston Chronicle 1995. "Big Changes in Big D," March 19, S1.

Horvath, R. 1971. "The 'Detroit Geographical Expeditions and Institute' Experience," *Antipode* 3, 73–85.

Horvath, R. 1974. "Machine Space," *Geographical Review* 64, 167–188.

Howell, P. 1993. "Public Space and the Public Sphere: Political Theory and the Historical Geography of Modernity," *Environment and Planning D: Society and Space* 11, 303–322.

Howland, G. 1994. "The New Outlaws: Cities Make Homelessness a Crime," *The Progressive* 58 (May), 33–35.

Hubbard, P. 1998. "Sexuality, Immorality and the City: Red-Light Districts and the Marginalisation of Female Street Prostitutes," *Gender, Place, and Culture* 5, 55–76.

Hubbard, P. 2001. "Sex Zones: Intimacy, Citizenship, and Public Space," *Sexualities* 4, 51–71.

HUD (U.S. Department of Housing and Urban Development). 1999. *Waiting in Vain: An Update on America's Rental Housing Crisis*. Washington, DC: Government Printing Office.

Industrial Relations. 1916. *Industrial Relations: Final Report and Testimony Submitted to Congress by the Commission on Industrial Relations*. Washington, DC: Senate Document 415, 64th Congress, 1st Session, Volume 11.

IWW. 1990. *IWW Songs We Never Forget*. Chicago: Industrial Workers of the World.

Jackson, K. 1985. *Crabgrass Frontier: The Suburbanization of the United States*. New York: Oxford University Press.

Jacobs, J. 1961. *The Life and Death of Great American Cities*. New York: Random House.

Jappé, A. 1999. *Guy Debord*. Berkeley: University of California Press (translated by D. Nicholson-Smith).

Jencks, C. 1981. *The Language of Postmodernism*. New York: Rizzoli (3rd edition).

Johnson, K. 1990. "Santa Ana Homeless Reject Deal in Sweep Case; Law Enforcement: A Jury Trial Becomes More Likely as Indigent Defendants Refuse the City's Offer to Have Them Plead Guilty and Pay Fines or Spend a Day in Jail," *Los Angeles Times*, October 20, B12.

Johnson, B., and Norse, R. 1996. "Marathon Protest Defies Santa Cruz Sleeping Ban." *Street Spirit* 2 (August), 1–11.

Kahn, B. 1991a. "People's Park: Is the Fight Over?" *East Bay Express*, March 22, 2, 28.

Kahn, B. 1991b. "Activists and Homeless Haggle Over Future of People's Park," *East Bay Express*, June 14, 3, 29–30.

Kahn, B. 1991c. "Who's in Charge Here? University Bulldozer Rolls While Council Is Out of Town," *East Bay Express* August 9, 1, 11–13.

Kalven, H. 1965. "The Concept of the Public Forum: Cox v. Louisiana," in P. Kurland (ed.), *The Supreme Court Review*. Chicago: Chicago University Press, 1–32.

Karacas, C. 2000. "A Fight for the Soul of the Haight: One Neighborhood's Trial by Space," Unpublished Paper, Department of Geography, University of California, Berkeley.

Kasinitz, P. 1986. "Gentrification and Homelessness: The Single Room Occupant and the Inner City Revival," in J. Erickson and C. Wilhelm (eds.), *Housing the Homeless*. New York: Center for Urban Policy Research, 241–252.

Katz, C. 1998. "Whose Nature, Whose Culture? Private Productions of Space and the 'Preservation of Nature,' " in B. Braun and N. Castree (eds.), *Remaking Reality: Nature at the Millennium*. New York: Routledge, 46–63.

Katz, C. 2001. "Hiding the Target: Social Reproduction in the Privatized Urban Environment," in C. Minca (ed.), *Postmodern Geography: Theory and Praxis*. Oxford: Blackwell, 94–110.

Kelley, R. 1998. *Yo' Mama's Disfunktional! Fighting the Culture Wars in Urban America*. Boston: Beacon.

Kelling, G. 1987. "Acquiring a Taste for Order: The Community and the Police," *Crime and Delinquency* 33, 90–102.

Kelling, G., and Coles, C. 1996. *Fixing Broken Windows: Restoring Order and Reducing Crime in Our Communities*. New York: The Free Press.

Kerr, C. 2001 (1963). *The Uses of the University*. Cambridge, MA: Harvard University Press.

Kerr, C., Dunlop, J., Harbison, F., and Myers, C. 1960. *Industrialism and Industrial Man*. Cambridge, MA: Harvard University Press.

Kilian, T. 1998. "Public and Private, Power and Space," in A. Light and J. Smith (eds.), *The Production of Public Space*. Lanham, MD: Rowman and Littlefield (Philosophy and Geography II, 115–134).

Kirsch, S. 1995. "The Incredible Shrinking World: Technology and the Production of Space," *Environment and Planning D: Society and Space* 13, 529–555.

Klein, N. 1999. *No Logo: Taking Aim at the Brand Bullies*. New York: Picador.

Kolodny, A. 1975. *The Lay of the Land: Metaphor as Experience and History in American Life and Letters*. Chapel Hill: University of North Carolina Press.

Koopman, J. 1991. "People's Park Protestors Brace for Today," *Contra Costa Times*, August 3, A1, A13.

Kowinski, W. 1985. *The Malling of America: An Inside Look at the Great Consumer Paradise*. New York: William Morrow.

Laclau, E., and Mouffe, C. 1985. *Hegemony and Socialist Strategy*. London: Verso.

Lee, H. 2000. "Poll Backs People's Park As Is, UC Berkeley Students Reject Alternative Use," *San Francisco Chronicle*, April 22, A13.

Lees, L. 1998. "Urban Renaissance and the Street: Spaces of Control and Contestation," in N. Fyfe, (ed.), *Images of the Street: Planning, Identity, and Control in Public Space*. London: Routledge, 236–253.

Lees, L. 2001. "Towards a Critical Geography of Architecture: The Case of an Ersatz Colosseum," *Ecumene* 8, 51–86.

Lefebvre, H. 1991 (1974). *The Production of Space*. Oxford: Blackwell (translated by D. Nicholson-Smith).

Lefebvre, H. 1996 (1968). "The Right to the City," in *Writing on Cities*. Oxford: Blackwell (edited and translated by E. Kofman and E. Lebas), 63–181, originally published as *Le Droit à la Ville*. Paris: Anthropos.

Leo, J. 1993. "Distorting the Homeless Debate," *U.S. News and World Report* 115 (November 8), 27.

Leonard, P., Dolbere, C., and Lazere, E. 1989. *A Place to Call Home: The Crisis in Housing for the Poor.* Washington DC: Center on Budget and Policy Priorities, Low Income Housing Information Center.

Levine, H. 1987. "Homeless Shelter Closes," *San Francisco Examiner* May 11, C1.

LFC (La Follette Commission) 1938. *Hearings*, Exhibit 9527, Gibson, Dunn and Crutcher to Bishop, August 26, 1938, United States Senate, Subcommittee of the Committee on Education and Labor, *Hearings on S. Res. 266: Violations of Free Speech and the Rights of Labor* (75 Parts), Part 61. Washington, DC: U.S. Government Printing Office, 22374–5.

LFC (La Follette Commission) 1944. *Violations of Free Speech and the Rights of Labor: Report of the Committee on Education and Labor Pursuant to S. Res. 266*, Report 398, Part IX. Washington, DC: U.S. Government Printing Office.

Lipietz, A. 1986. "New Tendencies in the International Division of Labor: Regimes of Accumulation and Modes of Regulation," in A. Scott and M. Storper (eds.), *Production, Work, Territory*. Boston: Allen and Unwin, 16–40.

Lipset, S. 1965. "University Student Politics," in S. Lipset and S. Wolin (eds.), *The Berkeley Student Revolt*. Garden City: Doubleday, 1–9.

Lipset, S., and Wolin, S. (eds.) 1965. *The Berkeley Student Revolt*. Garden City: Doubleday.

Longan, M. 2000. *Community and Place in Cyberspace: The Community Networking Movement in the United States*, Unpublished PhD Dissertation, Department of Geography, University of Colorado.

Los Angeles Times 1969. "Reagan Charges Park Riots Were Planned," May 21, 123.

Los Angeles Times 1987. "Original 'Skid Row' Homeless Add Sad Note to Gentrified Seattle Area," March 24, 1.20.

Los Angeles Times 1988. "Eviction of Homeless in Berkeley Sparks Melee," March 17, 13.

Los Angeles Times 1989a. "Rally at Berkeley Erupts into Riot," May 21, 13.

Los Angeles Times 1989b. "S.F. Clears Park's Tent City of Structures, Not People," July 21, 13.

Los Angeles Times 1990. "Compassion for the Homeless Wearing Thin in Bay Area," July 20, A1.

Los Angeles Times 1991a. "Berkeley Bastion," March 13, A3.

Los Angeles Times 1991b. "Temper Tantrums Over Dystopian Nightmare," August 7, A10.

Los Angeles Times 1992. "Play Replaces Protest at People's Park," March 31, A3.

Los Angeles Times 2001. "Secret Cameras Scanned Crowd at Super Bowl for Criminals," February 1, A2.

Low, S. 2000. *On the Plaza: The Politics of Public Space and Culture.* Austin: University of Texas Press.

Lowenthal, D. 1985. *The Past Is a Foreign Country.* Cambridge: Cambridge University Press.

Lukács, G. 1968. "Reification and the Consciousness of the Proletariat," in *History and Class Consciousness: Studies in Marxist Dialectics.* Cambridge, MA: MIT Press, 88–222.

Lyford, J. 1982. *The Berkeley Archipelago.* Chicago: Regnery Gateway.

Lynch, A. 1991a. "Council Recess Adds to 'People's Park' Woes," *San Francisco Chronicle* August 6, B1.

Lynch, A. 1991b. "Police Arrest Protesters at New Volleyball Courts," *San Francisco Chronicle* August 2, A1, A20.

Lynch, A., and Dietz, D. 1991. "Fewer Recruits for People's Park Wars," *San Francisco Chronicle* August 9, A1, A20.

Lyotard, J.-F. 1985. *The Postmodern Condition.* Minneapolis: University of Minnesota Press.

MacDonald, H. 1995. "San Francisco's Matrix Program for the Homeless," *Criminal Justice Ethics* 14(2), 79–80.

MacKenzie, E. 1994. *Privatopia: Homeowners Associations and the Rise of Residential Private Government.* New Haven: Yale University Press.

MacPherson, C. 1978. *Property: Mainstream and Critical Positions.* Toronto: University of Toronto Press.

McCann, E. 1999. "Race, Protest and Place: Contextualizing Lefebvre in the US City," *Antipode* 31, 163–184.

McCarthy, C. 1994. "Law vs. the Homeless," *Washington Post* December 27, 17.

McSheehy, W. 1979. *Skid Row.* Boston: G. K. Hall and Cambridge, MA: Schenkman Publishing.

McWilliams, C. 1971. *Factories in the Field.* Santa Barbara: Peregrine Smith. Originally published Boston: Little, Brown 1939.

Magnet, M. (ed.) 2000. *The Millennial City: A New Urban Paradigm for 21st-Century America.* Chicago: Ivan R. Dee.

Mair, A. 1986. "The Homeless and the Post-Industrial City," *Political Geography Quarterly* 5, 351–368.

Marcuse, P. 1988. "Neutralizing Homelessness," *Socialist Review* 18, 69–86.

Marston, S. 1990. "Who Are 'the People'? Gender, Citizenship, and the Making of the American Nation," *Environment and Planning D: Society and Space* 8, 449–458.

Marx, K. 1987. *Capital,* Volume 1. New York: International Publishers. Originally published 1867

Massey, D. 1995. *Space, Place, and Gender.* Minneapolis: University of Minnesota Press.

May, M. 1993. "Telegraph Ave. Shoppers Report Retail Revival," *The Bay Guardian,* January 9, 9.

Merrifield, A. 1993. "Place and Space: A Lefebvrian Reconciliation," *Transactions of the Institute of British Geographers,* 18, 516–531.

Merrifield, A. 1995. "Situated Knowledge Through Exploration: Reflections on Bunge's 'Geographical Expeditions,' " *Antipode* 27, 49–70.

Merrifield, A. 2002. *MetroMarxism.* New York: Routledge.

Merrifield, A., and Swyngedouw, E. 1996. "Social Justice and the Urban Experience," in A. Merrifield and E. Swyngedouw (eds.), *The Urbanization of Injustice.* New York: New York University Press, 1–17.

Meszeros, I. 1995. *Beyond Capital.* New York: Monthly Review Press.

Miller, D., Jackson, P., Thrift, N., Holbrook, B., and Rowlands, M. 1998. *Shopping, Place and Identity.* New York: Routledge.

Miller, J. 1994. *Democracy Is in the Streets.* Cambridge: Harvard University Press.

Miller, K. 2007. *Designs on the Public: The Private Lives of New York's Public Spaces.* Minneapolis: University of Minnesota Press.

Millich, N. 1994. "Compassion Fatigue and the First Amendment: Are the Homeless Constitutional Castaways?" *U.C. Davis Law Review* 27, 255–355.

Minnesota Geography Reading Group. 1992. "Collective Response: Social Justice, Difference and the City," *Environment and Planning D: Society and Space* 10, 589–595.

Mitchell, D. 1992. "Iconography and Locational Conflict from the Underside: Free Speech, People's Park, and the Politics of Homelessness in Berkeley, California," *Political Geography* 11, 152–169.

Mitchell, D. 1995. "The End of Public Space? People's Park, Definitions of the Public, and Democracy," *Annals of the Association of American Geographers* 85, 108–133.

Mitchell, D. 1996a. *The Lie of the Land: Migrant Workers and the California Landscape.* Minneapolis: University of Minnesota Press.

Mitchell, D. 1996b. "Political Violence, Order, and the Legal Construction of Public Space: Power and the Public Forum Doctrine," *Urban Geography* 17, 158–178.

Mitchell, D. 1997a. "State Restructuring and the Importance of 'Rights Talk,' " in L. Staeheli, J. Kodras, and C. Flint (eds.), *State Devolution in America: Implications for a Diverse Society.* Thousand Oaks, CA: Sage Publications, 7–38.

Mitchell, D. 1997b. "The Annihilation of Space by Law: The Roots and Implications of Anti-Homeless Laws in the United State," *Antipode* 29, 303–235.

Mitchell, D. 1998a. "Anti-Homeless Laws and Public Space I: Begging and the First Amendment," *Urban Geography* 19, 6–11.

Mitchell, D. 1998b. "Anti-Homeless Laws and Public Space II: Further Constitutional Issues," *Urban Geography* 19, 98–104.

Mitchell, D. 1998c. "The Scales of Justice: Localist Ideology, Large-Scale Production and Agricultural Labor's Geography of Resistance in 1930s California," in A. Herod (ed.), *Organizing the Landscape: Geographical Perspectives on Labor Unionism*. Minneapolis: University of Minnesota Press, 159–194.

Mitchell, D. 2000. *Cultural Geography: A Critical Introduction*. Oxford: Blackwell.

Mitchell, D. 2001a. "The Devil's Arm: Points of Passage, Networks of Violence and the Political Economy of Landscape," *New Formations* 43, 44–60.

Mitchell, D. 2001b. "Postmodern Geographical Praxis? The Postmodern Impulse and the War Against the Homeless in the Post-Justice City," in C. Minca (ed.), *Postmodern Geography: Theory and Praxis* (Oxford: Blackwell), 57–92.

Mitchell, D. 2002. "Controlling Space, Controlling Scale: Migrant Labor, Free Speech, and the Regional Development in the American West in the Early 20th Century," *Journal of Historical Geography*, 28, 63–84.

Mitchell, D. 2003. "California Living, California Dying: Dead Labor and the Political Economy of Landscape," in K. Anderson, S. Pile, and N. Thrift (eds.), *Handbook of Cultural Geography*. London: Sage, 233–248.

Mitchell, D. 2013a. "The Liberalization of Free Speech: How Protest Is Silenced in Public Space," in W. Nichols, B. Miller, and J. Beaumont (eds.), *Spaces of Contention: Spatialities and Social Movements*. Farnham, UK: Ashgate, 47–67.

Mitchell, D. 2013b. "Tent Cities: Interstitial Spaces of Survival," in A. Mubi Brighenti (ed.), *Urban Interstices: The Aesthetics and Politics of Spatial In-Betweens*. Farnham, UK: Ashgate, 65–85.

Mitchell, D., and Heynen, N. 2009. "The Geography of Survival and the Right to the City: Speculations on Surveillance, Legal Innovation, and the Criminalization of Intervention," *Urban Geography* 30, 611–632.

Mitchell, D., and Staeheli, L. 2002. "Clean and Safe? Property Redevelopment, Public Space, and Homelessness in Downtown San Diego." Paper presented at the Politics of Public Space Conference, Graduate Center, City University of New York, March 1.

Mitchell, D., and Van Deusen, D. 2002. "Downsview Park: A Missed Opportunity for a Truly Public Space?," in J. Czerniak (ed.), *CASE: Downsview Park*. Cambridge, MA: Harvard School of Design/Prestel Publishers, 102–113.

Mitchell, W. 1995. *City of Bits: Space, Place and the Infobahn*. Cambridge, MA: MIT Press.

Molotch, H. 1976. "The City as Growth Machine," *American Journal of Sociology* 82, 309–332.

Morgan, E. 1988. *Inventing the People: The Rise of Popular Sovereignty in England and America*. New York: W. W. Norton.

Mouffe, C. 1992. *Dimensions of Radical Democracy: Pluralism, Citizenship, Community*. London: Verso.

New York Times 1987. "Panhandling Law in Use in Seattle," 1987, 1.44.

New York Times 1988a. "A Playground Derelicts Can't Enter," August 20, A31.

New York Times 1988b. "29 Trying to Feed the Homeless Are Arrested in San Francisco," August 30, A14.

New York Times 1989. "Violence Flares at Berkeley Park During Event Marking 60's Battle," May 21, 1:26.

New York Times 1990. "Subway Panhandlers See Little From Legal Victory," January 29, B1.

New York Times 1991a. "Deal Is Struck on Fate of Park and Protest Site," March 10, 1:39.

New York Times 1991b. "Idealism to Decay to Volleyball at People's Park," July 5, A8.

New York Times 2001. "The Going Rate: 25 Cents," 2001, 4.2.

NLCHP (National Law Center on Homelessness and Poverty). 1995. *No Room for the Inn: A Report on Local Opposition to Housing and Social Services Facilities for Homeless People in 36 United States Cities*. Washington DC: National Law Center on Homelessness and Poverty.

NLCHP (National Law Center on Homelessness and Poverty). 1997. *Access Delayed, Access Denied: Local Opposition to Housing and Services for Homeless People Across the United States*. Washington DC: National Law Center on Homelessness and Poverty.

NLCHP (National Law Center on Homelessness and Poverty) 1999. *Out of Sight—Out of Mind?* Washington DC: National Law Center on Homelessness and Poverty.

Oc, T., and Tiesdale, S. 2000. "Urban Design Approaches to Safer City Centres: The Fortress, the Panoptic, the Regulatory and the Animated," in J. Gold and G. Revill (eds.), *Landscapes of Defence*. Harlow, UK: Pearson Education, 188–208.

Ollman, B. 1990. *Dialectical Investigations*. New York: Routledge.

Paisner, S. 1994. "Compassion, Politics, and the Problem Lying on Our Sidewalks: A Legislative Approach for Cities to Address Homelessness," *Temple Law Review* 67, 1259–1305.

Parker, C. 1919. *The Casual Laborer and Other Essays.* New York: Harcourt, Brace and Howe.

Pateman, C. 1989. *The Disorder of Women: Democracy, Feminism and Political Theory.* Stanford: Stanford University Press.

Patterson, R. 2012. "UC Must Transform People's Park's Legacy," *The Daily Californian,* February 3, 2012.

Peck, J. 1996. *Work-Place.* New York: Guilford Press.

Piven, F., and Cloward, R. 1992. *Regulating the Poor: The Functions of Public Welfare.* New York: Vintage (updated, revised edition).

Preston, W. 1963. *Aliens and Dissenters: Federal Suppression of Radicals, 1903–1933.* New York: Harper and Row.

Pue, W. 1990. "Wrestling with Law: (Geographical) Specificity vs. (Legal) Abstraction," *Urban Geography* 11, 556–585.

Pulido, L. 2000. "Rethinking Environmental Racism: White Privilege and Urban Development in Southern California," *Annals of the Association of American Geographers* 90, 12–40.

Rabinowitz, J. 1989. "Berkeley Journal: People's Park Struggle Resumes After 20 Years," *New York Times* April 24.

Rahimian, A., Wolch, J., and Koegel, P. 1992. "A Model of Homeless Migration: Homeless Men in Skid Row, Los Angeles," *Environment and Planning A* 24, 1317–1336.

Raspberry, W. 1992. "Telling the Truth About Homelessness," *Washington Post,* December 29.

Readings, B. 1996. *The University in Ruins.* Cambridge, MA: Harvard University Press.

Ribton-Turner, C. 1887. *A History of Vagrants and Vagrancy and Beggars and Begging.* London: Chapman Hall.

Rivlin, G. 1991a. "People's Park: Construction Zone," *East Bay Express* August 2, 3, 27.

Rivlin, G. 1991b. "Appropriate Force? Reports of Police Inflicted Injuries Continue to Flow In." *East Bay Express,* August 9, 1, 13–18.

Roberts, C. 1994. "Girding the Globe: The Boundaries Between People and Countries Are Being Erased by Telecommunications." *Boulder Daily Camera* February 10, C1.

Rodgers, T. 1992. "Many Panhandlers Not Homeless," *San Diego Union-Tribune* March 17, B1.

Roll Call 1993. "In the Neighborhoods: Shortchanged by the City's New Panhandling Law," June 17, n.p.

Rorty, R. 1996. "What's Wrong with 'Rights,' " *Harper's Magazine* 202, June, 15–18.

Rosati, C. 2002. "Humiliation for Accumulation's Sake." Paper presented at the

3rd International Critical Geography Conference, Békéscsaba, Hungary, June 29.

Rose, C. 1994. *Property and Persuasion: Essays on the History, Theory and Rhetoric of Ownership*. Boulder: Westview Press.

Rosen, J. 2001. "A Watchful State," *New York Times Magazine* October 7. Electronic version accessed through *www.nytimes.com*, available as of October 8, 2002.

Rosenberg, N. 1989. "Another History of Free Speech: The 1920s and the 1940s," *Law and Inequality* 7, 333–366.

Rossi, P. 1989. *Down and Out in America: The Origins of Homelessness*. Chicago: University of Chicago Press.

Rowe, S., and Wolch, J. 1990. "Social Networks in Time and Space: Homeless Women in Skid Row, Los Angeles," *Annals of the Association of American Geographers* 80, 184–204.

Ruddick, S. 1990. "Heterotopias of the Homeless: Strategies and Tactics of Placemaking in Los Angeles," *Strategies: A Journal of Theory, Culture, and Politics* 3, 184–201.

Ruddick, S. 1996. *Young and Homeless in Hollywood*. New York: Routledge.

Rule, J. 1988. *Theories of Civil Violence*. Berkeley: University of California Press.

Sack, R. 1986. *Human Territoriality*. Cambridge: Cambridge University Press.

Sampson, R., and Raudenbush, S. 2001. *Disorder in Neighborhoods: Does It Lead to Crime?* Washington: National Institute of Justice, Report NCJ 186049.

San Francisco Chronicle 1993. "Homeless Deaths Drop a Bit, But Still Top 100," December 21, A19.

San Francisco Chronicle 1994a. "All Sides Ready to Sound Off on Begging Law," February 15, A13.

San Francisco Chronicle 1994b. "Santa Cruz Wants to License Beggars," February 10, A21

San Francisco Chronicle 1997. "S.F. Reports Big Drop in Homeless Deaths," December 18, A21.

San Francisco Chronicle 1998. "Homeless Deaths in S.F. Are at Record High: Weather, Drug Abuse, Shortage of Rooms Are Possible Factors," December 12, A22.

San Francisco Chronicle 1999. "Record Number of Homeless People Died on S.F. Streets in '99," December 23, A24.

San Francisco Chronicle 2000. "Commemorating Homeless Deaths in One County," November 14, A22.

Scheer, R. 1969. "The Dialectics of Confrontation: Who Ripped Off the Park?" *Ramparts* 8 (August), 42–53.

Schein, R. 1997. "The Place of Landscape: A Conceptual Framework for an American Scene," *Annals of the Association of American Geographers* 87, 660–680.

Schneider, J. 1986. "Skid Row as an Urban Neighborhood, 1880–1960," in J. Erickson and C. Wilhelm (eds.), *Housing the Homeless*. New Brunswick: Center for Urban Policy Research, 67–89.

Schwantes, C. 1985. *Coxey's Army: An American Odyssey*. Lincoln: University of Nebraska Press.

Schwartz, B., and Kurtzman, L. 1988. "Tough New Homeless Policy Triggers Furor in Santa Ana," *Los Angeles Times*, July 10, 132.

Seattle Times 1993a. "Homeless Hit Streets to Protest Proposed Ban," August 28, A9.

Seattle Times 1993b. "Sidran Details Proposals to Control Street People," August 3, B1.

Seattle Times 1993c. "Will Tougher Panhandling Laws Work?" October 1, A1.

Sennett, R. 1992. *The Fall of Public Man*. New York: W. W. Norton.

Sennett, R. 1994. *Flesh and Stone: The Body and the City in Western Civilization*. New York: W. W. Norton.

Shields, R. 1998. *Lefebvre, Love and Struggle*. New York: Routledge.

Shklar, J. 1991. *American Citizenship: The Quest for Inclusion*. Cambridge, MA: MIT Press.

Sibley, D. 1995. *Geographies of Exclusion*. London: Routledge.

Sidran, M. 1993. "This Is the Best of Times to Keep This City Livable," *The Seattle Times*, August 10, B5.

Simon, H. 1992. "Towns Without Pity: A Constitutional and Historical Analyses of Official Efforts to Drive Homeless People from American Cities," *Tulane Law Review* 66, 631–676.

Simon, H. 1995. "The Criminalization of Homelessness in Santa Ana, California: A Case Study," *Clearinghouse Review* 29, 725–729.

Singer, D. 1999. *Whose Millennium? Theirs or Ours?* New York: Monthly Review Press.

Smith, D. 1994a. "A Theoretical and Legal Challenge to Homeless Criminalization as Public Policy," *Yale Law and Policy Review* 12, 487–517.

Smith, D. 1994b. *Geography and Social Justice*. Oxford: Blackwell.

Smith, I. 1996. "Arresting the Homeless for Sleeping in Public: A Paradigm for Expanding the Robinson Doctrine," *Columbia Journal of Law and Social Problems* 29, 293–335.

Smith, N. 1989. "Tompkins Square Park," *The Portable Lower East Side* 6(2), 1–28.

Smith, N. 1990. *Uneven Development: Nature, Capital, and the Production of Space*. Oxford: Blackwell (2nd edition).

Smith, N. 1992a. "Contours of a Spatialized Politics: Homeless Vehicles and the Production of Geographical Scale," *Social Text* 33, 55–81.

Smith, N. 1992b. "New City, New Frontier: The Lower East Side as Wild, Wild, West," in M. Sorkin (ed.), *Variations on a Theme Park: The New American City and the End of Public Space*. New York: Hill and Wang, 61–93.

Smith, N. 1993. "Homeless/Global: Scaling Places," in J. Bird, B. Curtis, T. Putnam, G. Robertson, and L. Tickner (eds.), *Mapping the Futures: Local Cultures, Global Change*. London: Routledge, 87–119.

Smith, N. 1996. *The New Urban Frontier: Gentrification and the Revanchist City*. New York: Routledge.

Smith, N. 1998. "Giuliani Time: The Revanchist 1990s," *Social Text* 57, 1–20.

Smith, N. 2000. "The Restructuring of Spatial Scale and the New Global Geography of Uneven Development," *Jimbun Chiri* 52 (1), 51–65.

Smolla, R. 1992. *Free Speech in an Open Society*. New York: Alfred A. Knopf.

Soja, E. 1989. *Postmodern Geographies: The Reassertion of Space in Critical Social Theory*. Oxford: Blackwell.

Soja, E. 1996. *ThirdSpace*. Oxford: Blackwell.

Solenberger, A. 1911. *One Thousand Homeless Men*. New York: Russell Sage.

Sorkin, M. (ed.) 1992. *Variations on a Theme Park: The New American City and the End of Public Space*. New York: Hill and Wang.

Spradley, J. 1970. *You Owe Yourself a Drunk: An Ethnography of Urban Nomads*. Boston: Little, Brown.

Staeheli, L. 1994. "Restructuring Citizenship in Pueblo, Colorado." *Environment and Planning A* 26, 849–871.

Staeheli, L. 1996. "Publicity, Privacy, and Women's Political Action," *Environment and Planning D: Society and Space* 14, 601–619.

Staeheli, L., and Cope, M. 1994. "Empowering Women's Citizenship," *Political Geography* 13, 443–460.

Staeheli, L., and Mitchell, D. 2008. *The People's Property?: Power, Politics, and the Public*. New York: Routledge.

Stern, S. 1987. "Activists Seek a Solution Beyond Shelters," *Oakland Tribune*, February 22, D10.

Storper, M., and Walker, R. 1989. *The Capitalist Imperative: Territory, Technology and Industrial Growth*. Oxford: Blackwell.

Street, S. 2001. "Promoting Order or Squelching Dissent? Protesters and Civil Libertarians Object to the Use of 'Free-Speech Zones,' " *Chronicle of Higher Education*, January 12, A37–A38.

Sunstein, C. 1992. "Free Speech Now," in G. Stone, R. Epstein, and C. Sunstein (eds.), *The Bill of Rights in the Modern State*. Chicago: University of Chicago Press.

Takahashi, L. 1996. "A Decade of Understanding Homelessness in the USA: From Characterization to Representation," *Progress in Human Geography* 20, 291–310.

Takahashi, L. 1998. *Homelessness, AIDS, and Stigmatization: The NIMBY Syndrome in the United States at the End of the Twentieth Century*. Oxford: Oxford University Press.

Tier, R. 1993. "Maintaining Safety and Civility in Public Spaces: A Constitutional Approach to Aggressive Begging," *Louisiana Law Review* 54, 285–338.

Tier, R. 1998. "Restoring Order in Urban Public Spaces," *Texas Review of Law and Politics* 2, 256–291.

Tomasky, M. 1995. *Left for Dead: The Life, Death, and Possible Resurrection of Progressive Politics in America.* New York: Free Press.

Tushnet, M. 1984. "An Essay on Rights," *Texas Law Review* 62, 1363–1412.

Van Deusen, R. 2002. "Urban Design and the Production of Public Space in Syracuse, New York," Paper Presented at the Rights to the City Conference, Rome, June 1.

Veness, A. 1993. "Neither Homed Nor Homeless: Contested Definitions and Personal Worlds of the Poor," *Political Geography* 12, 319–340.

Vidler, A. 2001. "Aftermath: A City Transformed: Designing 'Defensible Space,' " *New York Times*, September 23, 4:6.

Waldron, J. 1991. "Homelessness and the Issue of Freedom," *UCLA Law Review* 39, 295–324.

Walker, R. 1996. "Another Round of Globalization in San Francisco," *Urban Geography* 17, 60–94.

Walker, S. 1990. *In Defense of American Liberties: A History of the ACLU.* New York: Oxford University Press.

Wallace, M. 1996. *Mickey Mouse History and Other Essays on American Memory.* Philadelphia: Temple University Press.

Wallace, S. 1965. *Skid Row as a Way of Life.* Totowa, NJ: The Bedminster Press.

Walters, P. 1990. "Orange County Voices; Fixing Public's 'Broken Windows'; Santa Ana Civic Center Disorder is a 'Broken Window' That Invites More Antisocial Behavior If Not Fixed By Police," *Los Angeles Times*, August 28, B9.

Walzer, M. 1986. "Public Space: Pleasures and Costs of Urbanity," *Dissent* 33, 470–475.

Warren, S. 1994. "Disneyfication of the Metropolis: Popular Resistance in Seattle," *Journal of Urban Affairs* 16, 89–107.

Washington Post 1993a. "D.C. Officers Skeptical of Panhandling Limits," June 3, B1.

Washington Post 1993b. "Panhandlers Tap Deep Pockets of Resentment," May 9, B1.

Weinstock, H. 1912. *Report of Harris Weinstock, Commission to Investigate Recent Disturbances in the City of San Diego and the County of San Diego, California to His Excellency Hiram W. Johnson, Governor of California.* Sacramento: State Printing Office.

Wells, J. 1994. "Begging Law Endorsed by Berkeley Council, Mayor Proposes Ordinance as Part of Compromise," *San Francisco Chronicle*, June 15, A14.

Will, G. 1987. "Living on the Street: Mentally Ill Homeless Contribute to Community Decay," Syndicated Column, Washington Post Writers' Group.

Will, G. 1995. "A Court Victory for Public Order," Syndicated Column, Washington Post Writers' Group.

Will, G. 1997. "Liberalism's Urban Ruins," Syndicated Column, Washington Post Writers' Group.

Williams, K., Johnstone, C., and Goodwin, M. 2000. "CCTV Surveillance in Urban Britain: Beyond the Rhetc.   f Crime Prevention," in J. Gold and G. Revill (eds.), *Landscapes of Defer.*   Harlow, UK: Pearson Education, 168–187.

Williams, P. 1991. *Alchemy of Race and Rights.* Cai.   dge, MA: Harvard University Press.

Williams, R. 1977. *Marxism and Literature.* Oxford: Oxi.   University Press.

Williams, R. 1983. *Keywords.* London: Fontana Press.

Williams, R. 1997. *Problems of Materialism and Culture.* L   on: Verso. Originally published 1980.

Wilson, A. 1992. *The Culture of Nature: From Disney to the Exxon Va...* Oxford: Blackwell.

Wilson, D. 1991. "Urban Change, Circuits of Capital, and Uneven Development," *Professional Geographer* 43, 403–415.

Wilson, E. 1991. *The Sphinx in the City: Urban Life, the Control of Disorder, and Women.* Berkeley: University of California Press.

Wilson, J. 1968. *Varieties of Police Behavior.* Cambridge: Cambridge University Press.

Wilson, J. 1996. "Forward," in G. Kelling and S. Coles, *Fixing Broken Windows.* New York: The Free Press, xiii–xvi.

Wilson, J., and Kelling, G. 1982. "Broken Windows: The Police and Neighborhood Safety," *Atlantic Monthly* (March), 29–38.

Winerip, M. 1999. "Bedlam on the Streets: Increasingly the Mentally Ill Have No Place to Go," *New York Times Magazine,* May 23, 42–49, 56, 65–66, 70.

Wolch, J. 1980. "The Residential Location of the Service-Dependent Poor," *Annals of the Association of American Geographers* 70, 330–341.

Wolch, J., and Dear, M. 1993. *Malign Neglect: Homelessness in an American City.* San Francisco: Jossey-Bass.

Wolch, J., Rahimian, A, and Koegel, P. 1993. "Daily and Periodic Mobility Patterns of the Urban Homeless," *Professional Geographer* 45, 159–169.

Wolff, R. 1966. "Letter to the Editor," *Atlantic Monthly* 218, November, 38.

Wong, E. 1999. "California and the West; A '60's-Style Uproar Over People's Park; Berkeley: A Suggestion to Build Dorms on the Site of Counterculture Clashes Ignites Some of the Same Passions That Raged 30 Years Ago," *Los Angeles Times* May 9, A22.

Young, I. 1990. *Justice and the Politics of Difference.* Princeton: Princeton University Press.

Zukin, S. 1991. *Landscapes of Power: From Detroit to Disney World*. Berkeley: University of California Press.

Zukin, S. 1995. *The Cultures of Cities*. Oxford: Blackwell.

CASES CITED

Abrams v. United States 1919. 250 US 616.

American Steel Foundries v. Tri-Cities Central Trades Council 1921. 257 US 184.

Atchison, Topeka and Santa Fe Railway v. Gee 1905. 139 F. 582 (CCSD, Iowa).

Carlson v. California 1940. 310 US 746.

CBS v. DNC 1973. 412 US 94.

Clark v. Community for Creative Non-Violence 1984. 468 US 288.

Debs v. United States 1919. 249 US 211.

DeJong v. Oregon 1937. 299 US 353.

Frohwerk v. United States 1919. 249 US 204

Fox v. Washington 1915. 236 US 273.

Gitlow v. New York 1925. 263 US 252.

Hague v. CIO 1939. 307 US 496.

Hill et al. v. Colorado et al. 2000. 120 S. Ct. 2480.

In re Eichorn 1998. 81 Cal. Rptr. 2d 535 (Ct. App).

In re Phelan 1894 (*Thomas v. Cincinnati, N.O. & T.P. Railway*) 62 7d. 803 (CCSD, Ohio).

Johnson v. City of Dallas 1994. 860 F. Supp. 344 (N.D. Tex.)

Joyce v. San Francisco 1994. 846 F. Supp. 843 (N.D. Cal.)

Lamont v. Postmaster General 1965. 381 US 301.

Madsen v. Women's Health Center 1994. 114 S. Ct. 2516.

Operation Rescue v. Women's Health Center, Inc. 1993. 626 So. 2d. 664.

Patterson v. Colorado 1907. 205 US 454.

Pottinger v. City of Miami 1994. 810 F. Supp. 1551 (S.D. Fla.)

Robinson v. California 1962. 370 US 660.

Schench v. United States 1919. 249 US 47

Thornhill v. Alabama 1940. 310 US 88.

Tobe v. Santa Ana 1994. 27 Cal Rpte. 2d 386 (Cal. Ct. App.)

Tobe v. Santa Ana 1995. 892 P. 2d. 599 (Cal.)

Truax v. Carrigan 1921. 257 US 312.

United States v. Kokinda 1990. 110 S. Ct. 3115.

Vegelahn v. Guntner 1896. 167 Mass. 92, 44 N.E. 1077.

Index

Q

R